Debt Dance: A Global Journey Through Fiscal Tightropes

Rany

Table of Contents

Chapter 3: What drives the Commodity-Sovereign Risk Dependence in Emerging Market Economies?   87

Chapter 4: Financial Linkages and Sectoral Business Cycle Synchronization: Evidence from Europe   145

Chapter 5: Physical Climate Change Risks and the Sovereign Creditworthiness of Emerging Economies   182

Chapter 1:

Introduction

1.1 Motivation: The Curious Case and Multiple Facets of Sovereign Debt

The fiscal costs of a country's public debt are of crucial importance to all governments. Any unit of currency that is spent to service interest payments on government bonds is a unit of currency that cannot be used to finance other expenditures a government may wish to pursue. Should public debt costs become unsustainable, a country might have to default on its debt obligations. Between 1970 and 2017, 75 episodes of sovereign debt restructuring occurred globally, which often entail substantial economic costs (Borensztein & Panizza 2009, Laeven & Valencia 2018). During the start of the Corona-Pandemic in 2020, so far 102 countries applied for emergency financing from the IMF to manage their debt (IMF 2020*b*).

Public debt is an issue with many facets. The controversy starts by arguing whether the economic costs of public debt outweigh its benefits and how much debt is sustainable, a discussion that has concerned economists for generations.[1]

Given that nearly all countries in the world are indebted, a first glance at public debt can sometimes reveal curious economic characteristics: Whereas Japan's public debt to GDP ratio of roughly 235% in 2019 seems to be manageable for the country, the same debt ratio would likely lead to over-indebtedness and sovereign default for most other nations.[2] Furthermore, while Japan's last primary fiscal surplus dates back to 1992, Italy has since this time almost always reported positive primary surpluses, but has nevertheless found its debt sustainability often times in question. Meanwhile, Ireland had a record fiscal deficit of 32% to GDP in 2010, which was mainly used to bail out the country's ailing banking sector – a move that bankrupted Ireland in the process. Finally, whereas rising refinancing costs for Greece almost pushed the country out of the Eurozone after 2010, Germany has since 2016 experienced episodes of negative interest rates for newly issued 10-year government bonds – that is, investors were willing to pay to enter the "safe haven" of German government debt.

---

[1]For Adam Smith, government bonds constituted a transfer of the productive group of capitalists to the rather unproductive group of financial creditors (Smith 1776 [2010]). For Karl Marx, government bonds were fictitious capital, as it generated returns without being profitable on its own (Marx 1894 [1992]). For a more modern assessment, see Blanchard (2019)'s AEA presidential address, arguing that the fiscal costs of debt are lower if safe interest rates are below economic growth rates.

[2]All data in this paragraph comes from the IMF Fiscal Monitor.

This book will focus on the effects of three key factors – bank distress, commodity prices and climate change – on the return dynamics of government bonds traded on financial markets. An additional chapter demonstrates the importance of financial market integration for the international propagation of business cycles. Together, these four chapters can shed some light on the "curious characteristics" of government debt and its policy implications.

With an outstanding volume of $45 trillion of marketable central government debt in OECD countries, the sovereign bond market accounts for a large share of global financial assets (OECD 2019). Though the market for government debt is therefore deep and liquid, it is also subject to change and disruptions. Government bonds of economically developed countries, once considered to be a nearly risk-free investment, have come under increased scrutiny by market participants. The number of sovereign borrowers with AAA-rating by Standard & Poor's has dropped from 20 to only 12 between 2007 and 2017, as the financial and Eurozone crisis have revealed flaws in the assessment of government bonds as near risk-free (Financial Times 2017). By contrast, the number of countries rated below investment grade has risen from 49 to 62. During the Corona-Pandemic, even the market for US Treasuries, the world's supposedly most liquid and safest financial asset, experienced severe disruptions as uncertainty about the pandemic spread (Brookings Institution 2020).

Further controversy comes from ownership questions of sovereign debt. As noted by Arbogast (2020), a disaggregated ownership structure of public debt holdings, mapping which households, banks and firms hold which amount of sovereign bonds, is difficult to come by, even for official debt management offices. This issue can gain political traction, for instance if the broader public has to bear the costs of government expenditure cuts implemented to service a group of anonymous bondholders.

In general, having a more international investor base is associated with lower sovereign bond costs and therefore targeted by public debt managers (Abbas et al. 2014). However, foreign investors can also be a less stable source of funding that withdraws at the first signs of economic stress, leading to spiking debt costs (Arslanalp & Tsuda 2014). What for some is the "enforcement of market discipline" by international investors that require compensation for their risk-taking, is for others "bond-market terror" by "freebooting market vigilantes" (Tooze 2020).

For the Eurozone, we know from aggregated data that financial integration up to 2007 led to a more international investor base of government bond holdings (Bruegel 2019). However, this trend reversed during the Eurozone crisis, as banks stepped in as buyers of sovereign debt of their home country, particularly in crisis-hit countries, while international investors withdrew cross-border holdings. A "doom loop" between weak sovereigns and banks emerged. Consequently, several reform proposals intending to improve the Eurozone architecture aim at reducing the home bias in sovereign bond exposures of banks through regulatory interventions (Bénassy-Quéré et al. 2018). Given that, on the other hand, governments tended to perform moral suasion on their banks to purchase government bonds, thus stabilizing yields, the issue of exposure limits has led to fierce p olitical d ebates (Altavilla e t a l. 2017).[3] I n s um, the sovereign-bank loop became a major determinant in the integration process of the Eurozone and will be the central theme of the second chapter of this book.

Whereas debt costs are important but often manageable for highly-rated countries, debt burdens can be much more crucial issues for emerging economies. High debt costs bind fiscal r esources t hat c ould o therwise b e u sed t o fi ght po verty or fo ster he alth an d education systems. Debt reliefs for low-income countries are therefore frequently debated, and as of 1996 managed by the HIPC initiative of the IMF and World Bank (IMF 2020a).

In this context, the ability to service government bond obligations is not only a function of fiscal e xpenditure r isks, b ut a lso o f t he s tability a nd s trength o f g overnment revenues. Countries can become dependent on key revenue sources, which can lead to political and economic difficulties if the corresponding income streams are outside of government control like in the case of commodities.

According to UNCTAD (2019), 102 out of 189 countries worldwide were considered "commodity-dependent" between 2013 and 2017, implying that major parts of their exports constitute of raw materials. Crucially, 91% of low-income countries fall under this classification (UNCTAD 2019). Higher commodity prices have historically often shaped the boom-bust dynamics of countries, leading first t o c apital i nflows, st ronger fis cal pos itions and expenditures, but triggering crises and sovereign defaults once raw material prices decline (Reinhart et al. 2016). The discussions on debt standstills or debt reliefs for low-income countries are

---

[3] In a panel discussion at the Halle Institute for Economic Research in 2020, one Member of the European Parliament described the subject of sovereign bond exposure limits for banks in negotiations as "He who must not be named", referencing the Harry Potter book series, to highlight the political sensitivity of the issue.

a direct consequence of their commodity-based economic structure (United Nations 2020). Hence, Chapter 3 will investigate the implications of commodity prices on sovereign bond conditions for emerging economies in greater detail.

The analysis of sovereign debt is also related to the integration of financial systems. As financial markets can propagate business cycle swings, the success of policy responses towards a common shock critically hinges on the degree of integration between financial systems. National policies alone might be insufficient to fight common cyclical shocks or market-wide financial crises. For economically integrated regions such as Europe, supranational approaches might be a more promising way to prevent business cycle downturns.

For instance, one of the most important supranational measures against business cycle downturns during the Euro and the Corona-crisis was the ECB's monetary policy. In general, central banks in advanced economies have stepped up as major buyers of sovereign debt through their asset purchase programs known as quantitative easing. The ECB has acquired Euro Area government bonds worth over 2,000 EUR billion with its public sector purchase program between 2015 and 2020 and announced further bond purchases through its pandemic emergency purchase program in 2020 of so far 1,350 EUR billion (ECB 2020a,b).

In the context of the Euro crisis, ECB President Mario Draghi's 2012 commitment to do "whatever it takes" was primarily targeted at easing turmoil and lowering premia in the market for sovereign bonds (Draghi 2012). His announcement calmed bond markets and is publically often attributed to saving the Euro (Financial Times 2019). Thus, a supranational response was necessary to address a business cycle downturn in a region of highly integrated financial markets. The importance of business cycle swings in integrated financial systems will be the central theme of Chapter 4 of this book.

A further topic deals with the repayment of public debt. Government bonds, unlike traditional loans, are not secured by collateral or other legal enforcement measures like covenants. In contrast to corporate bonds, asset holders have much more limited claims towards a defaulting sovereign than a defaulting firm (Panizza et al. 2009). As Eaton & Gersovitz (1981) point out, nothing prevents governments from defaulting on their debt, except for the damage to their reputation that would exclude them from further borrowing. Nevertheless, the literature argues that sovereign debt contracts, especially for emerging economies, require sanctions and legal enforcement to make sure countries repay their creditors (Bulow & Rogoff 1989). However, legal frameworks also have to apply for creditors, as certain investors could otherwise engage in "holdout" strategies, that is, they refuse to engage in restructuring the debt of a defaulting sovereign and demand full repayment. Today, these politically highly sensitive legal issues for countries unable to repay their debt are often managed by "collective action clauses" attached to sovereign bonds (Panizza et al. 2009).

A novel source of legal questions concerning sovereign bond repayment is posed by climate change. Governments are increasingly facing legal consequences for not disclosing climate risks in their sovereign bond disclosures (Bloomberg 2020). As global warming and natural disasters can lead to a deterioration in the market price of sovereign debt for countries, rating agencies have started incorporating climate change predictions into their sovereign rating (Moody's 2016). Chapter 5 of this book will analyze the nexus between climate change and sovereign creditworthiness in greater detail.

All these observations, though they are partially anecdotal, illustrate the importance to understand the factors that shape the risks, fiscal costs and sustainability of sovereign debt, as these issues entail significant implications for monetary policy, economic prosperity and political stability. What ultimately decides if a country can manage its debt is whether it can find buyers on financial markets to finance new government bonds, and on the financial conditions they demand.[4] Higher interest rates on government bonds can endanger the solvency of a sovereign, hence the term *"sovereign risk"*. Since maturing government debt is rarely repaid permanently, but typically rolled-over by issuing new bonds, the analyzed re-financing costs are a crucial link between a country's accumulated debt, its fiscal situation and current bond market conditions and therefore of central importance for all governments (Buchheit 2020). To identify the economic factors on financial markets that affect sovereign risk, their economic importance and the channels through which they drive government borrowing costs is the central research contribution of this book.

---

[4] Assuming no direct sovereign financing via central banks.

1.2   Outline and Contribution of this book to the Literature

While the previous observations on the distribution, level, benefits o r c osts o f p ublic debt sketched out the field on the analysis of sovereign risk, the following section describes in detail the central contributions of this book.

The presented chapters apply panel regressions to empirically investigate key determinants of sovereign risk. Higher sovereign risk (or declining sovereign creditworthiness), which is defined as increasing sovereign bond interest rates, implies that it is more expensive for governments to issue bonds. Previous research has shown that sovereign risk is to large degrees driven by global factors such as US stock returns or the volatility index VIX (Longstaff et al. 2011). Nevertheless, even after controlling for global factors, country-specific macroeconomic conditions such as terms of trade variation, trade openness or fiscal b alances c an a ffect the sovereign risk level to a critical degree (Aizenman et al. 2016, Edwards 1984, Hilscher & Nosbusch 2010). Furthermore, sovereign creditworthiness not only responds to economic news but is also driven by a country's political environment (Eichler 2014). Also, sovereign bond pricing can have psychological elements, as investors can suddenly become more sensitive towards a country's economic fundamentals during a crisis, experiencing a "wake-up call" (Beirne & Fratzscher 2013). Lastly, sovereign bond spreads might also spike excessively beyond fundamental considerations, especially if financial m arkets a re i n s tates o f p anic and overly pessimistic (Aizenman et al. 2013).

Chapter 2 of this book adds to the literature by improving the empirical identification of one of the key drivers of sovereign risk in the Euro Area, i.e. the sovereign-bank loop. The corresponding paper is co-authored with Stefan Eichler and published in the Journal of Financial Stability. Bank distress and sovereign distress can reinforce each other, for instance if expensive bank bailouts weaken the fiscal c apacity o f t he g overnment. T he economic identification of the risk transmissions between banks and sovereigns in the previous literature is therefore subject to reverse causality and other endogeneity concerns such as unobserved Eurozone exit risks (e.g. Acharya et al. (2014), Altavilla et al. (2017)).

The chapter contributes to the literature by isolating the bank-to-sovereign distress channel within the sovereign-bank loop, thus overcoming reverse causality effects. To do so, we collect stock prices of 132 Eurozone banks from 1999 to 2016 for nine Eurozone member countries. We subsequently build asset-weighted bank stock returns for each country that serve as bank distress measures. In the following, we instrument bank distress using non-Eurozone stock market returns that are weighted according to the claims of the Eurozone banking sector towards the respective non-Eurozone country. These exposure shocks, approximating Eurozone bank losses or asset write-downs in foreign markets, are unlikely to be affected by domestic sovereign risk variation. We take additional steps to address potentially endogenous disturbances in the identification by using fixed bank claim data or dropping non-Eurozone markets that depend heavily on credit from the Euro Area. The resulting foreign exposure shocks prove to be a strong instrument for Eurozone bank distress.

With this novel identification at hand, we estimate a statistically significant transfer of instrumented bank distress on sovereign distress. Bank risk was therefore one of the major causes of the Eurozone crisis, not just a by-product or correlation, which is an important finding for the literature and policy debate. However, the isolated bank-to-sovereign risk transmission is significantly smaller than the corresponding OLS regression. This result suggests that reverse causality and other biases are indeed present in the unadjusted estimation of the sovereign-bank loop.

In an appendix to the second chapter, we also estimate the drivers of the bank-to-sovereign risk transmission. The corresponding specification refers specifically to the Eurozone crisis from 2009 to 2016. We find that countries with less stable macroeconomic conditions, impaired financial sectors and higher political uncertainty feature a statistically significantly stronger transfer of bank-to-sovereign distress. All these results are important contributions to the literature and the political debate on a more sustainable Eurozone architecture.

Chapter 3 of the book analyses the role of international commodity prices as one of the key determinants of sovereign risk for emerging economies. It is co-authored with Stefan Eichler and Stefan Giessler. While the importance of commodity prices for business cycles is well documented (see Fernández et al. (2018)), less is known about their impact on sovereign risk and in particular the drivers of the commodity-sovereign risk relationship. We therefore contribute to the literature, first, by building a portfolio of weighted commodity price returns, in which a country's commodity export shares form the respective weights. This measure on its own is statistically significant in explaining sovereign risk, which we measure by daily changes in the Emerging Market Bond Index Spread (EMBI) for a panel of 34 emerging economies from 1994 to 2016. The effect also holds when we control for exporters with world-market power, when taking imported commodities into account or when measuring export performance as an index instead of a return variable.

Using interaction models, we observe that countries with greater commodity export shares have a stronger effect of commodity prices on their sovereign risk level. Next to this intuitive result, we find that energy commodities lead to stronger, while exporting industrial metals to a weaker commodity dependence. Other commodity-specific characteristics, such as the price volatility of the commodities or their diversification within the portfolio are of less importance. Furthermore, we uncover that commodity dependence increases in times of recessions and when public and private sectors lack fiscal resources. Also, countries with a more stable repayment history of sovereign debt are subject to lower commodity dependence.

Turning to the policy measures that could mitigate commodity dependence, our results show that improving institutional quality is associated with statistically significantly lower effects of commodity prices on sovereign risk. Better institutions could therefore be important by providing clear ownership rights in the extraction process and by spending the revenues from commodity exports efficiently. In addition, fostering manufacturing sectors, investments in physical capital and infrastructure, attracting FDI inflows, opening trade accounts and accumulating international reserves are all strongly associated with lower commodity dependence. All these results are important extensions to the literature and provide guidance for policy makers.

Chapter 4 investigates the effects of financial market integration on business cycle synchronization in Europe. Co-authored with Julia Schaumburg and Lena Tonzer, the paper addresses the question if more financial integration leads to converging business cycles (such as in Imbs (2006)) or diverging business cycle patterns (such as in Kalemli-Ozcan et al. (2013)). The former explanation could be at work if financial markets transmit shocks across a network of integrated countries, while the latter effect could emerge when a recession in one country causes lower bank lending to local firms, as integrated financial markets redirect funding towards economies that are not in crisis.

We collect data on financial integration, measured by cross-border bank claims, for 10 European countries from 1996 to 2017. Our contribution to the literature consists of a dynamic spatial network approach. Following Blasques et al. (2016), we construct financial linkages between countries as a matrix that reflects time-varying economic distances. Financial integration among countries is thus not only measured between country-pairs, as in the previous literature, but for a network of countries. In this way, we take endogenous feedback effects of countries' GDP fluctuations into account. A second advantage is that the dynamic modeling structure estimates a time-varying parameter that approximates the sign and strength of business cycle co-movement that is due to financial integration and can thus shed a light on the conflicting result in the previous literature.

We start by investigating the spillover degree of GDP growth for our panel countries. The dynamic parameter shows graphically that financially more integrated countries tend to have on average positive business cycle synchronization. However, the effect size depends crucially on time and industry. During crises periods, such as the financial or the Euro crisis, financial integration leads to much higher co-movements of European business cycles than in calmer times. Furthermore, by disentangling overall GDP growth into the value-added of different industries, we show that business-sensitive sectors such as wholesale & retail trade or industrial production are driving these results. On the other hand, industries like construction or agriculture have roughly constant and rather small spillover effects over time. Industries that are largely nationally regulated such as public administration or arts & entertainment do not show any signs of dynamic spillovers in our model, which is a reasonable result.

Since we observe on average converging business cycle effects due to financial integration, in particular in times of crisis, our results suggest that focusing on national approaches to stabilize business cycles has likely limited effects. Supranational policy measures, such as the European Banking Union, are thus an important addition to deal with international shocks.

The final Chapter 5 contributes to a novel strand of literature. In this single-authored paper, I show that climate change risks can impair the sovereign creditworthiness of emerging economies. In face of rising temperatures due to climate change, further warming and increasing weather irregularities can likely lead to a deterioration in the market price of sovereign debt for countries that are susceptible to these trends (Bank of England 2018).

In this chapter, I link the literature on temperature fluctuations, which shows that higher temperatures lead to lower GDP growth, firm productivity and political stability (Burke et al. 2015, Dell et al. 2012, Hsiang et al. 2013, Zhang et al. 2018) with the literature on sovereign bond performance. I draw long-term monthly temperature data since 1901 for a panel of

54 emerging economies. Next, I calculate the deviation of a country's observed temperature since 1994 from the historical average of temperature of this country between 1901 and 1950. This measure approximates the degree of global warming experienced so far, and averages around 0.84°C which is close to the estimations of the IPCC (2018).

I subsequently regress EMBI data on a monthly frequency from 1994 to 2018 on a country's temperature deviation. My results indicate that temperature increases can statistically significantly impair sovereign creditworthiness, but only for susceptible countries. More specifically, I find that countries that already have a warm climate suffer significantly more from further warming with respect to their sovereign risk level. In addition, countries with a low quality of institutions bear statistically significantly larger sovereign risk burdens from rising temperatures than nations with better institutions. Stronger institutional quality is therefore important, for instance when swift and efficient decisions are required after a natural disaster or to handle the distributional consequences of climate change. Importantly, the institution-channel and the warmness-channel hold simultaneously, which implies that all countries can adapt to climate change risks by improving their institutional quality even after controlling for their general warmness.

In sum, my methodological contribution is, first, to investigate novel channels of sovereign risk. The impact of banking distress on sovereign risk (Chapter 2) has only gained traction since the financial and Euro crisis. Chapter 5, analyzing the effects of physical climate risks on sovereign risk, is indeed one of the first papers in this field.

Second, I strive for consistent and if possible exogenous effects by testing sovereign risk determinants that are outside of government control, such as temperature fluctuations (Chapter 5) or international commodity prices (Chapter 3). In case of the sovereign-bank relationship in Chapter 2, an instrumental variable approach to isolate the global variation in bank risk that is outside of government control is applied. I do not claim to derive causal effects like in quasi-natural experiments, as these are rare on a country-level. Nevertheless, my empirical specifications address the major threats to a clear statistical inference by appropriate control variables and fixed effects, alternative specifications and several robustness tests.

Third, the effort for stronger identification is enhanced by applying high-frequency daily or monthly financial market data. Other than macroeconomic statistics, finance data is published without delay. Furthermore, it can shed some light on the underlying causes of sovereign bond turbulences. For instance, using direct stock data of Irish banks in 2010 can give a more timely and precise picture to describe Irish sovereign risk than the resulting 32% to GDP fiscal deficit mentioned in the introduction. Moreover, higher data frequencies reduce endogeneity concerns further, as financial market participants have to treat, for instance, the institutional quality of a country on a given day as fixed, other than in yearly or quarterly regressions. Lastly, using financial market data motivates the fourth chapter of this book which deals with the synchronization of business cycle fluctuations due to financial integration.

I derive policy contributions by testing a comprehensive set of macroeconomic, financial, or institutional channels for the respective sovereign risk relationship, which is informative for policy makers. For instance, in Chapter 3, I find that the impact of a country's commodity performance on sovereign risk can be mitigated if the country improves its institutional quality and fosters downstream production technologies. The methodology of interaction regressions used to derive these results have, if correctly applied as described by Brambor et al. (2006), strong statistical properties that allow for consistent inferences (Bun & Harrison 2019, Nizalova & Murtazashvili 2016).

In sum, this book attempts to present consistent and well identified empirical estimations, that analyze meaningful determinants of sovereign risk, while investigating which politically relevant channels drive the respective sovereign risk connection.

A.1   Appendix to Chapter 1

In accordance with §10 (3) 4-6 PromO2018, I will describe in the following my individual contributions to all chapters of this book.

For Chapter 2, the main research idea was developed jointly with my co-author Stefan Eichler. While Stefan Eichler had the initial idea to investigate country-specific bank distress in the Euro Area in the spirit of Buch & Neugebauer (2011), we both recognized the importance to apply an instrumental variable approach and that foreign exposure shocks could be an appropriate candidate. I further worked on strengthening the instrument's exogeneity, by applying fixed exposure weights, dropping the most credit-dependent Eurozone borrowers and by orthogonalizing the instrument with respect to Eurozone-specific variation. We both agreed to use sovereign bond spreads towards Germany as our dependent variable.

Moreover, I collected all data on sovereign bond spreads, bank stock returns, banks' foreign exposures as well as financial market control variables. Additionally, I conducted all regressions, both for the benchmark and for the channels of the bank-to-sovereign distress. Therefore, I was responsible for the specific regression design in terms of fixed effects, winsorizing of variables or the choice of control variables, which was finalized in consultation with Stefan Eichler. We jointly decided to focus on the size difference between the IV and the OLS estimation as one of the key contributions of the paper. While Stefan Eichler guided the process, I was responsible for working out the different channels of the bank-sovereign relationship, and grouping them into macroeconomic, government bond, banking sector and political categories. In addition, I tested several methodological approaches suggested by Stefan Eichler on how to best capture the determinants of bank-to-sovereign distress, for instance, by drawing the quarterly betas of the benchmark regression and applying them as a new dependent variable. In the end, we jointly decided to apply an interaction model as the most appropriate method. I also developed all robustness tests, for instance to use bank-specific shocks as an alternative instrument, since this variation is less affected by trade shocks or other outside influences. Furthermore, I wrote the first draft of the paper including tables, graphs and the literature review, to which Stefan Eichler provided valuable comments and suggestions and re-drafted certain parts.

Regarding Chapter 3, which is co-authored with Stefan Eichler and Stefan Giessler, the paper's main idea to test for the effect of commodity prices on sovereign risk, by applying a country-specific export-weighted commodity portfolio, was already formulated by Stefan Eichler. However, I contributed to the project's idea, first, by strengthening the exogeneity of the baseline effect. I suggested and carried out to lag the export weights by one year to avoid multicollinearity with the commodity prices, to test for the impact of commodity exporters with world market power and to take the effect of commodity imports into account. I also implemented a test in which I transformed the commodity performance variable into an index in the spirit of a difference-in-difference model. While Stefan Giessler collected data on the commodity exports of every country, my second major contribution to the project was the collection of all remaining data, i.e. EMBI spreads, commodity prices, control variables and conditioning variables used to describe the channels of the commodity-sovereign risk relationship. I also matched the commodity export products with the best available corresponding commodity price and calculated the resulting commodity portfolios. Third, I estimated all regressions in the paper and tested for different channels driving our benchmark. I also came up with the idea to test the diversification of the commodity portfolio and its standard deviation of prices as possible drivers. With Stefan Eichler providing feedback, we jointly decided to group the chapters on the estimated channels into commodity-related, macroeconomic and policy-related subsections. Fourth, I drafted the first version of the paper including all tables and graphs, which was supplemented and re-drafted in certain parts by Stefan Eichler. I conducted an extensive literature review, to which Stefan Eichler and Stefan Giessler both added several papers to the final version. Also, Stefan Giessler carried out a different method to estimate the channels of the commodity-sovereign risk relationship, which we, however, did not use in the final paper.

For Chapter 4, co-authored with Lena Tonzer and Julia Schaumburg, both co-authors had developed a broad research outline of the paper when I joined the project. The paper should be an application of Blasques et al. (2016), which is co-authored by Julia Schaumburg, with respect to financial integration. I started contributing to the project, first, by conducting an extensive literature review. By doing so, we could work out the conflicting effects of financial integration on business cycle synchronization in previous papers. Furthermore, my review of the literature gave us insights on the most appropriate data sources for our paper, for instance to use the locational banking statistics by the Bank for International Settlement as a measure for financial i ntegration. Lastly, my overview showed us that, next to the methodological side, a focus on industrial sector production was, with the exception of one paper, so far lacking in the literature.

While the model specification o f t he d ynamic s patial a pproach w as h andled b y Julia Schaumburg, my second major contribution to the project was the collection of data. I drew data on financial i ntegration, i ndustrial value a dded a nd s everal c ontrol v ariables such as productivity growth from different sources. In addition, I constructed the time-varying matrices that measured financial integration in our network of countries which would directly feed into Julia Schaumburg's spatial model. Third, I applied Julia Schaumburg's code to estimate the dynamic spatial model. In doing so, I also included the final s et o f control variables, which was decided jointly with Lena Tonzer. I tested a vast amount of model specifications, for which I d iscussed t he r esults w ith L ena Tonzer a nd t he t echnical d etails of the model with Julia Schaumburg, who subsequently adjusted the code to determine the most appropriate specification. We j ointly d iscussed t he fi nal re sults in cluded in th e pa per, while the economic interpretation and the embedding of our paper in the literature were headed by Lena Tonzer and me.

The writing process was shared among all authors, with Lena Tonzer initiating a first draft. I added several tables of data, results and summary statistics to the paper, as well as descriptions on the data sources and several comments and redrafts to improve the paper, which marks my fourth contribution to the project. Further result tables and graphs on the estimated spillover parameters as well as the model description were provided by Julia Schaumburg.

Chapter 5 of this **book** is single-authored. Hence, all data collections, estimations and drafting choices were handled by me. Nevertheless, I would like to point out that I also developed the research question on climate risks and sovereign creditworthiness on my own. All people who provided valuable feedback to the paper are named in the first f ootnote of Chapter 5.

References to Chapter 1

Abbas, S. A., Blattner, L., De Broeck, M., El-Ganainy, M. A. & Hu, M. (2014), 'Sovereign debt composition in advanced economies: A historical perspective'. International Monetary Fund WP/14/162.

Acharya, V., Drechsler, I. & Schnabl, P. (2014), 'A pyrrhic victory? Bank bailouts and sovereign credit risk', *The Journal of Finance* **69**(6), 2689–2739.

Aizenman, J., Hutchison, M. & Jinjarak, Y. (2013), 'What is the risk of European sovereign debt defaults? Fiscal space, CDS spreads and market pricing of risk', *Journal of International Money and Finance* **34**, 37–59.

Aizenman, J., Jinjarak, Y. & Park, D. (2016), 'Fundamentals and sovereign risk of emerging markets', *Pacific Economic Review* **21**(2), 151–177.

Altavilla, C., Pagano, M. & Simonelli, S. (2017), 'Bank exposures and sovereign stress transmission', *Review of Finance* **21**(6), 2103–2139.

Arbogast, T. (2020), 'Who are these bond vigilantes anyway? The political economy of sovereign debt ownership in the Eurozone'. MPIfG Discussion Paper No. 20/2.

Arslanalp, S. & Tsuda, T. (2014), 'Tracking global demand for advanced economy sovereign debt', *IMF Economic Review* **62**(3), 430–464.

Bank of England (2018), 'Transition in thinking: The impact of climate change on the UK banking sector'.

Beirne, J. & Fratzscher, M. (2013), 'The pricing of sovereign risk and contagion during the European sovereign debt crisis', *Journal of International Money and Finance* **34**, 60–82.

Bénassy-Quéré, A., Brunnermeier, M., Enderlein, H., Farhi, E., Fuest, C., Gourinchas, P.-O., Martin, P., Pisani-Ferry, J., Rey, H., Schnabel, I. et al. (2018), 'Reconciling risk sharing with market discipline: A constructive approach to Euro Area reform'. CEPR Policy Insight No. 91.

Blanchard, O. (2019), 'Public debt and low interest rates', *American Economic Review* **109**(4), 1197–1229.

Blasques, F., Koopman, S. J., Lucas, A. & Schaumburg, J. (2016), 'Spillover dynamics for systemic risk measurement using spatial financial time series models', *Journal of Econometrics* **195**(2), 211–223.

Bloomberg (2020), 'Australia sued for not disclosing climate risk in sovereign debt'. `https://www.bloomberg.com/amp/news/articles/2020-07-22/australia-sued-for-not-disclosing-climate-risk-in-sovereign-debt?__twitter_impression=true`.

Borensztein, E. & Panizza, U. (2009), 'The costs of sovereign default', *IMF Staff Papers* **56**(4), 683–741.

Brambor, T., Clark, W. R. & Golder, M. (2006), 'Understanding interaction models: Improving empirical analyses', *Political Analysis* pp. 63–82.

Brookings Institution (2020), 'How did COVID-19 disrupt the market for U.S. Treasury debt?'. `https://www.brookings.edu/blog/up-front/2020/05/01/how-did-covid-19-disrupt-the-market-for-u-s-treasury-debt/`.

Bruegel (2019), 'Whose (fiscal) debt is it anyway?'. Bruegel Blog, `http://bruegel.org/2019/02/whose-fiscal-debt-is-it-anyway/`.

Buch, C. M. & Neugebauer, K. (2011), 'Bank-specific shocks and the real economy', *Journal of Banking & Finance* **35**(8), 2179–2187.

Buchheit, L. (2020), 'From coronavirus crisis to sovereign debt crisis'. Financial Times, `https://ftalphaville.ft.com/2020/03/25/1585171627000/From-coronavirus-crisis-to--sovereign-debt-crisis/`.

Bulow, J. I. & Rogoff, K. (1989), 'Sovereign debt: Is to forgive to forget?', *The American Economic Review* **79**(1), 43–50.

Bun, M. J. & Harrison, T. D. (2019), 'OLS and IV estimation of regression models including endogenous interaction terms', *Econometric Reviews* **38**(7), 814–827.

Burke, M., Hsiang, S. M. & Miguel, E. (2015), 'Global non-linear effect of temperature on economic production', *Nature* **527**(7577), 235–239.

Dell, M., Jones, B. F. & Olken, B. A. (2012), 'Temperature shocks and economic growth: Evidence from the last half century', *American Economic Journal: Macroeconomics* **4**(3), 66–95.

Draghi, M. (2012), 'Speech by Mario Draghi, President of the European Central Bank at the global investment conference in London 26 July 2012'. `https://www.ecb.europa.eu/press/key/date/2012/html/sp120726.en.html`.

Eaton, J. & Gersovitz, M. (1981), 'Debt with potential repudiation: Theoretical and empirical analysis', *The Review of Economic Studies* **48**(2), 289–309.

ECB (2020*a*), 'Asset purchase programmes'. `https://www.ecb.europa.eu/mopo/implement/omt/html/index.en.html#pspp`.

ECB (2020*b*), 'Pandemic emergency purchase programme (PEPP)'. `https://www.ecb.europa.eu/mopo/implement/pepp/html/index.en.html`.

Edwards, S. (1984), 'LDC's foreign borrowing and default risk: An empirical investigation', *The American Economic Review* **74**(4), 726–734.

Eichler, S. (2014), 'The political determinants of sovereign bond yield spreads', *Journal of International Money and Finance* **46**, 82–103.

Fernández, A., González, A. & Rodriguez, D. (2018), 'Sharing a ride on the commodities roller coaster: Common factors in business cycles of emerging economies', *Journal of International Economics* **111**, 99–121.

Financial Times (2017), 'Supranational debt issuance more than doubles in a decade'. `https://www.ft.com/content/481139b8-7c36-11e7-ab01-a13271d1ee9c`.

Financial Times (2019), 'Draghi's ECB tenure: Saving the euro, faltering on inflation'. `https://www.ft.com/content/a62b221c-eb64-11e9-a240-3b065ef5fc55`.

Hilscher, J. & Nosbusch, Y. (2010), 'Determinants of sovereign risk: Macroeconomic fundamentals and the pricing of sovereign debt', *Review of Finance* **14**(2), 235–262.

Hsiang, S. M., Burke, M. & Miguel, E. (2013), 'Quantifying the influence of climate on human conflict', *Science* **341**(6151), 1235367.

Imbs, J. (2006), 'The real effects of financial integration', *Journal of International Economics* **68**(2), 296–324.

IMF (2020*a*), 'Factsheet: Debt relief under the heavily indebted poor countries (hipc) initiative'. `https://www.imf.org/en/About/Factsheets/Sheets/2016/08/01/16/11/Debt-Relief-Under-the-Heavily-Indebted-Poor-Countries-Initiative`.

IMF (2020*b*), 'The IMF's response to COVID-19'. `https://www.imf.org/en/About/FAQ/imf-response-to-covid-19#Q1`.

IPCC (2018), 'Global Warming of 1.5°C. An IPCC Special Report on the impacts of global warming of 1.5°C above pre-industrial levels and related global greenhouse gas emission pathways, in the context of strengthening the global response to the threat of climate change, sustainable development, and efforts to eradicate poverty (summary for policymakers)', *Intergovernmental Panel on Climate Change* .

Kalemli-Ozcan, S., Papaioannou, E. & Peydró, J.-L. (2013), 'Financial regulation, financial globalization, and the synchronization of economic activity', *The Journal of Finance* **68**(3), 1179–1228.

Laeven, L. & Valencia, F. (2018), 'Systemic banking crises revisited'. IMF Working Paper 18/206.

Longstaff, F. A., Pan, J., Pedersen, L. H. & Singleton, K. J. (2011), 'How sovereign is sovereign credit risk?', *American Economic Journal: Macroeconomics* **3**(2), 75–103.

Marx, K. (1894 [1992]), *Capital: volume III*, Penguin UK.

Moody's (2016), 'Environmental risks - sovereigns: How Moody's assesses the physical effects of climate change on sovereign issuers'. `https://www.moodys.com/research/Moodys-sets-out-approach-to-assessing-the-credit-impact-of--PR_357629`.

Nizalova, O. Y. & Murtazashvili, I. (2016), 'Exogenous treatment and endogenous factors: Vanishing of omitted variable bias on the interaction term', *Journal of Econometric Methods* **5**(1), 71–77.

OECD (2019), 'Sovereign borrowing outlook for OECD countries'. OECD Sovereign Borrowing Outlook 2019.

Panizza, U., Sturzenegger, F. & Zettelmeyer, J. (2009), 'The economics and law of sovereign debt and default', *Journal of Economic Literature* **47**(3), 651–98.

Reinhart, C. M., Reinhart, V. & Trebesch, C. (2016), 'Global cycles: Capital flows, commodities, and sovereign defaults, 1815-2015', *American Economic Review:: Papers & Pro-

ceedings **106**(5), 574–80.

Smith, A. (1776 [2010]), *The Wealth of Nations: An inquiry into the nature and causes of the Wealth of Nations*, Harriman House Limited.

Tooze, A. (2020), 'Time to expose the reality of "debt market discipline"'. International Politics and Society, `https://www.ips-journal.eu/regions/europe/article/show/time-to-expose-the-reality-of-debt-market-discipline-4381/`.

UNCTAD (2019), 'The state of commodity dependence 2019'. United Nations conference on trade and development.

United Nations (2020), 'Commodity exporters face mounting economic challenges as pandemic spreads'. UN/DESA Policy Brief 60.

Zhang, P., Deschenes, O., Meng, K. & Zhang, J. (2018), 'Temperature effects on productivity and factor reallocation: Evidence from a half million Chinese manufacturing plants', *Journal of Environmental Economics and Management* **88**, 1–17.

Chapter 2:

Avoiding the Fall into the Loop: Isolating the Transmission of Bank-to-Sovereign Distress in the Euro Area

2.1 Introduction

The fatal relationship between bank and sovereign distress in the Eurozone, the *sovereign-bank loop*, has put the political and economic survivability of the currency union to the test. Bank distress started to amplify sovereign distress in particular with the beginning of the Eurozone crisis, as expensive financial sector bailouts weakened the fiscal capacity of sovereigns (*bailout channel*). In addition, deteriorating sovereign creditworthiness was transmitted to domestic banks' balance sheets through their vast holding of government debt securities. Impaired by these negative shocks, banks holding larger amounts of sovereign debt from GIIPS countries are associated with reducing credit supply, thus hampering general economic activity (*sovereign-bond channel*). Ultimately, distressed banks made further public rescue efforts more likely which once again endangered sovereign solvency. Taken together, both channels place the sovereign-bank loop as one of the primary reasons for the escalation of the Eurozone crisis as well as the sluggish economic recovery of the Eurozone after the financial crisis (Brunnermeier et al. 2016).

In light of these interlinkages, the identification of isolated channels within the sovereign-bank loop remains challenging given reverse causality between bailout and sovereign-bond channel and Eurozone-specific omitted variables such as Eurozone exit risks, implicit guarantees, political risks, and other institutional specifics. In this paper, we propose a novel instrumental variable approach to isolate the transmission of bank-to-sovereign distress within the loop, thereby addressing the endogenous nature of the sovereign-bank relationship.

To this end, we collect stock returns of 132 banks in the Eurozone to construct country-specific bank distress measures for nine members of the Euro Area. We instrument these bank returns on the country level using *non*-Eurozone stock market returns weighted with the bilateral claims of the Eurozone country's banking sector against the respective non-Eurozone country taken from the BIS Consolidated Banking Statistics. These imported exposure shocks indicate, if negative, loan losses or asset write-downs for banks in the respective market and thus drive stock returns of Eurozone banks. More importantly, our instrument is less likely to be affected by sovereign risk of the respective member state or Eurozone-specific unobservables since we exploit risks from outside the Eurozone (the weighted average of non-Eurozone returns) and take additional steps to remove any Eurozone related variation in them. More specifically, we use claim data from 1999:Q1 to rule out that Eurozone banking sectors endogenously shifted their international credit exposure as a response to the financial or Euro crisis. We also drop all non-Eurozone countries in the construction of the instrument if they depend heavily on financing from a Eurozone country and might therefore react more distinctly to Eurozone bank distress. Lastly, we orthogonalize the instrument with respect to Eurozone-specific variation of stock returns, thus removing any Euro-related correlation in the instrument. Our approach therefore isolates as precisely as possible the effects of non-Eurozone-driven exposure shocks affecting Eurozone banking sector distress which ultimately impacts sovereign distress.

One further potential concern for our identification strategy could be that non-Eurozone stock returns may have an effect on Eurozone sovereign risk that is not transmitted through the banking system, thereby violating the exclusion restriction of the instrument. For instance, the deterioration of the economic stance in a non-Eurozone country indicated by falling stock returns could be associated with worsening export opportunities and thus higher sovereign spreads of an Eurozone country. To remedy this concern, first, we control for non-financial stock market returns for each Eurozone country which should account for real economic shocks to the Eurozone country's economy, for instance due to deteriorating economic conditions of a non-Eurozone trading partner. Second, we control for the nominal effective exchange rate of the Euro, which should account for trade shocks affecting the Euro Area. We conduct further robustness tests to address this concern. First, we repeat our baseline estimation with non-Eurozone stock market returns consisting only of bank stocks as such bank-specific shocks are more likely to be transmitted only through the banking sector of a country, in line with the exclusion restriction. Second, we construct a measure of trade-specific shocks and add it as an additional control variable. Third, we drop countries with significant claims against the considered Eurozone country. For example, negative shocks in non-Eurozone countries with large holdings of Eurozone sovereign bonds may be associated with selling pressure and hence direct price losses on Eurozone sovereign bond prices. While this effect may not be entirely transmitted by Eurozone banks, excluding non-Eurozone countries with significant claims against the respective Eurozone country from our instrument should rule out this channel. All robustness tests confirm our main results. While we cannot categorically rule out further channels that could be a threat to the exclusion restriction, such as trade finance shocks, these robustness tests suggest that they are unlikely to disturb our main results.

We empirically test the hypothesis if bank distress instrumented by foreign exposure shocks is a statistically significant determinant of Eurozone sovereign risk. Our analysis uncovers an economically meaningful and highly statistically significant impact of the instrumented banking sector distress of a Euro Area country on its sovereign distress level. A one standard deviation increase in the instrumented national-specific bank distress of a Eurozone country leads to a rise in national sovereign distress by 0.0828 standard deviations on average. Banking sector distress was therefore a dominant factor for the propagation of the Eurozone crisis. However, the risk transmission coefficient from our instrumental variable (IV) regression is around 50% smaller than the one obtained using ordinary least squares (OLS) and both coefficients differ in a statistically significant manner. This result points to an aforementioned bias due to omitted variables and/or reverse causality which are unaddressed in the OLS framework. The gap between instrumented and non-instrumented coefficients remains when limiting the sample to the Eurozone crisis, using different dependent variables, measures for bank distress, data frequencies and other versions of the instrument.

Using our instrumental variable setting, we contribute to the literature by improving the identification of the transmission of bank-to-sovereign distress. Our results allow the conclusion that isolated bank distress is a dominant factor for the escalation and severity of the Eurozone crisis. Our analysis suggests that it is important to distinguish between the bailout and the sovereign bonds channel since both channels amplify each other when using OLS which leads to sizeable reverse causality. From a policy perspective, we provide evidence that banking sector stress is not just a by-product of the Eurozone crisis but one of the major factors or even potential causes for its propagation.

The rest of this article is organized as follows: In Section 2.2, we conduct a literature review on related articles and highlight the contribution of our paper. Section 2.3 describes the data we use and the construction of the instrument. Following on this, Section 2.4 presents our empirical strategy and compares results from the OLS framework to the IV estimation. We conduct encompassing robustness checks in Section 2.5. Section 2.6 concludes.

2.2 Literature Review

This article is related to a large body of literature that has studied the feedback loop between bank and sovereign distress during the Eurozone crisis. In their seminal paper, Acharya et al. (2014) show that with the onset of the Eurozone crisis in 2009-10 and the private-to-public risk transfers during the financial crisis, sovereign and banking distress started to positively intensify each other. This bailout-channel likely emerges from the fact that governments might face insolvency themselves if they allocate vast fiscal means for the rescue of an ailing and potentially too-big-to-fail banking sector. Alter & Schüler (2012) find results in accordance with this channel. Gerlach et al. (2010) research the determinants of rising sovereign bond spreads in the Euro Area and find that the balance sheet size of a country's banking sector relative to the country's GDP to be a significant determinant of rising sovereign bond spreads relative to Germany.

The sovereign-bond channel is constituted by several incentives banks face to hold domestic government debt. This effect is shaped by the zero risk weight of such assets in the calculation of banks' capital ratios (the risk shifting hypothesis, see Acharya & Steffen (2015),

Acharya, Eisert, Eufinger & Hirsch (2018), Buch et al. (2016), Kirschenmann et al. (2017)), political pressure by their home governments (the moral suasion hypothesis, see Becker & Ivashina (2017), De Marco & Macchiavelli (2016), Ongena et al. (2019)), monetary policy interventions (Crosignani et al. (2020), Drechsler et al. (2016)) or a combination of these factors (Altavilla et al. (2017), Horváth et al. (2015)). In the context of the sovereign-bank loop, banks' inclination towards government debt of their home state might stabilize their sovereign's bond spread as the domestic banking sector acts as a "buyer of last resort" for these securities (see Crosignani (2017)). However, the sovereign-bond-channel also constitutes a direct transmission of increasing sovereign distress to bank balance sheets which might impair banks' lending activities. In accordance with this channel, Popov & Van Horen (2015) and Acharya, Eisert, Eufinger & Hirsch (2018) find that banks that hold larger exposures of sovereign debt by GIIPS countries issue less credit to non-financial firms.

Our research is furthermore connected to Battistini et al. (2014) who split sovereign risk into a country-specific and a common component. The authors find evidence that banking sectors in periphery countries, but not core countries, respond to increases in the country-specific risk factor by expanding their holding of domestic sovereign debt. Comparably, we focus in our analysis on country-specific banking sector distress that captures idiosyncratic variation in the bank stock returns of a country. De Bruyckere et al. (2013) study the contagion between bank and sovereign risks during the Eurozone crisis and find that lower Tier 1 ratios on the banking and higher debt-to-GDP ratios on the country level are connected to stronger contagion. Schnabel & Schüwer (2016) investigate the relationship between financial and sovereign risk over time and find that the magnitude of the loop was largest in the period between 2010 and 2013 before it contracted somewhat while gaining new momentum in 2016. Fratzscher & Rieth (2019) highlight the two-way causality between banking and sovereign distress in a VAR-approach and find that the ECB's non-standard monetary policies and bank bailout announcements reduced credit risks of sovereigns and banks. Singh et al. (2016) also use bank distress measures based on stock market data to investigate the direction of bank-to-sovereign and sovereign-to-bank distress transfers and find evidence for both types of spillovers. Though they identify spillovers based on Granger causalities, both our paper and Singh et al. (2016) find that with the beginning of the financial and Euro crisis, the spillover direction from banks to sovereigns became more forceful. Nevertheless, as we show in Section

2.5.1, the difference between isolated bank-to-sovereign transfers and the unadjusted OLS framework remains substantial, even with the onset of the Euro crisis. Breckenfelder & Schwaab (2018) also recognize difficulties in handling the two-way dependence of bank and sovereign risks and use the ECB's stress test results from 2014 as a quasi-natural experiment to isolate bank-sovereign distress spillovers, finding that bank risks in stressed countries spilled over to non-stressed countries. Kallestrup et al. (2016) use a measure comparable to ours by multiplying the international exposures of a country's banking sector derived from BIS consolidated statistics with the banking or sovereign risk of the respective foreign country. However, they do not use this measure as an instrument for national bank distress, but rather to show correlations between domestic and foreign bank risk. Furthermore, and in contrast to Kallestrup et al. (2016), our focus is in particular on the sovereign-bank loop in the Euro Area as this currency union is special compared to countries with independent monetary policy and national currencies (see also De Grauwe & Ji (2013)). Despite these differences, the results between our paper and Kallestrup et al. (2016) are comparable: Whereas they report that an increase by one standard deviation in bank risk leads to a rise in the standard deviation of sovereign risk by 25%, this value is at 8.3% in our IV framework. This smaller magnitude is likely because we control for reverse causality factors and have a longer sample period starting in the (calmer) year 1999. In sum, we aim to contribute to the literature by proposing a novel instrumental variable setting to improve the identification of the transmission of bank-to-sovereign distress, thereby adding an additional layer of exogeneity to the above literature.

2.3 Data Description

2.3.1 Deriving Country-Specific Bank Distress

Our banking sector distress measure is based on daily stock returns of publicly-traded banks in the nine major Eurozone (EZ) countries[5] for the period of 1999-2016. Hence, we include all major events since the introduction of the Euro, such as the financial crisis and the Eurozone crisis. We include all frequently traded Eurozone banks available from Datastream during our observation period, arriving at 132 banks in total (see a list of the banks and their mean

---

[5]Austria, Belgium, France, Greece, Ireland, Italy, Netherlands, Portugal, Spain. We have to exclude Germany as it is the reference country in the construction of the dependent variable, i.e. sovereign return spreads. However, we include German banks for the construction of Eurozone-specific variables further below. We also disregard small countries such as Luxembourg or Malta and Finland, as they have both bad BIS and bank stock coverage.

asset size in the Appendix in Table A.2.7). The mean asset sizes of the banks indicate that we have financial institutions of various sizes. Given that our bank distress measure is based on asset weights, our results are driven by large international banks that are likely to respond to non-Eurozone exposure shocks as captured in our instrument.

Weighted bank stock returns are a simple measure for the fragility or distress of a country's banking sector. The bank stocks in our panel are frequently traded, making the stock price an easily observable and daily available measure. In the absence of new stock issuances or buy-backs, stock price movements correspond to changes in the market value of the equity of a bank. Falling stock prices can therefore be a clearer signal for a deterioration of a bank's fundamentals, as measured by the market, than balance sheet items which vary only quarterly and are subject to the reporting habits by the regulator.[6] It is also common in the financial literature to use buy-and-hold stock returns as a measure for a bank's performance over a certain period (see Beltratti & Stulz (2012), Fahlenbrach & Stulz (2011)) and it is intuitive to assume that a poor performance by a large, potentially too-big-to-fail bank has an adverse impact on sovereign creditworthiness. Lastly, stock market data has been frequently used to measure bank fragility in similar contexts (Bongini et al. (2002), Demirgüç-Kunt & Huizinga (2013), Eichler & Sobański (2016), Gropp et al. (2006)).

Though CDS spreads are also a common distress measure for banks, we choose stock returns, first, because they were more liquid during the crisis. Using CDS would force us to drop several Greek or Portuguese banks in the analysis as their CDS spreads turned illiquid. Second, stock returns are a better match for our instrument which relies on the international exposure of a banking sector towards all borrowers of a country, i.e. they encompass all sectors, whereas CDS spreads are sector-specific. Third, stock returns are more accurate to incorporate future profit expectations than CDS premiums. Imported stock return shocks as we construct them should therefore encompass international economic activity more broadly than CDS spreads.

We weight the daily stock return of each bank with its yearly total asset share on its home country level and then aggregate these weighted return series of a country's banking

---

[6]For instance, in September 2015 Moody's (2015) downgraded Greek banks to C despite their CET1 equity ratios being fairly above the regulatory requirement (e.g. 12.1% in the case of National Bank of Greece). However, a major part of this equity consisted of state preference shares and deferred tax assets which are considered to be low-quality equity. The banks' stock prices, on the other hand, had been declining for months at this point.

sector for each country i in our sample ($BankReturns_{it}$). This step provides us with a daily measure for bank distress on the country level in which the largest banks of the respective country have the greatest weight.[7]

However, the bank stock returns from which each country's return series is built will be subject to national, Eurozone and global variation in stock prices and economic activities. In the Eurozone, banking distress after the common shock of the financial crisis often took place on a national level, such as Ireland's bank bailout in 2010, Spain's nationalization of Bankia in 2012 or Italy's series of rescue packages for its ailing banking sector in 2015 and 2016. This national component of bank distress, separated from common financial distress affecting all EZ states at the same time, will likely react more strongly to country-specific macroeconomic or political factors. In order to construct a measure for bank distress on the country level that is only driven by national and global factors, i.e. to remove the Eurozone component in stock prices, we proceed as follows:

Similarly to the bank return series on the country level, we derive a Eurozone-specific return series by weighting the bank stock returns of our nine Eurozone sample countries and Germany with their asset share on the Eurozone level, i.e. of the asset size of all banks in our panel, and subsequently aggregate all weighted returns on the Euro Area level. Since this return series will still feature a global component of stock prices, we orthogonalize this Euro Area return series with respect to a Datastream bank stock return series for all global banks, except those from the Euro Area. Doing so, we clean the Eurozone bank returns from worldwide, non-Eurozone variation in stock prices and therefore isolate the Eurozone-specific component in bank stocks ($EurozoneBankReturns_t$).

Finally, we follow the approach of Buch & Neugebauer (2011) and subtract these Eurozone-specific bank returns from the derived bank stock return series for each Euro Area country i, arriving at a bank distress measure that picks up country-specific variations of bank distress, separated from common financial distress affecting all Eurozone states similarly.[8] We multi-

---

[7]We set the return of a bank to missing if the bank was delisted or taken over to control for survivorship bias. We also disregard stock returns when there was no turnover of the stock. If the stock return of a bank in a given quarter was missing for more than seven consecutive trading days, we set all returns of the bank in the respective quarter to missing in order to avoid jumps in the indices we construct in the following. This procedure affects mainly small banks with low trading volumes and is not critical for our results.

[8]We subtract the Eurozone-specific returns to be as close as possible to Buch & Neugebauer (2011). In a robustness check, we also remove Eurozone-specific returns by orthogonalization. Both distress measures correlate at almost 99% and we arrive at nearly identical results.

ply with minus one to interpret the returns as a distress measure, for which higher values of the distress measure indicate higher levels of distress.

$$\Delta NationalBankDistress_{it} = (-1) * (BankReturns_{it} - EurozoneBankReturns_t) \quad (1)$$

Though we receive similar results if we apply our following analysis to the unadjusted $BankReturns_{it}$, as shown in the robustness section, we have reason to believe that the national-specific bank distress measure approximates more precisely for the observed events of idiosyncratic bank distress, e.g. during the Eurozone crisis.

Table 2.1 provides summary statistics of ΔNationalBankDistress for the five GIIPS countries three months before different key events of banking sector turmoil during the crisis, showing in all cases considerable distress on average and elevated standard deviations.

Table 2.1: Summary statistics of country-specific bank distress three months before events of financial sector turmoil during the Eurozone crisis

ΔNationalBankDistress	Obs.	Mean	Median	Std. Dev.	Min	Max
Ireland: 09/16/2010–12/16/2010 (when EU-IMF bailout was signed)	66	0.604	1.096	4.132	-6.709	6.974
Spain: 03/09/2012–06/09/2012 (when EU-IMF bailout was agreed upon)	63	0.328	0.197	1.398	-2.363	4.306
Portugal: 05/03/2014–08/03/2014 (when bailout for Espirito Santo was announced)	65	0.322	0.576	2.597	-6.709	6.974
Italy: 04/29/2016–07/29/2016 (when ECB stress test results were announced in which Italian banks performed poorly)	66	0.238	0.462	2.506	-5.944	6.974
Greece: 01/26/2010–04/27/2010 (when Greece was downgraded to junk status)	61	0.384	0.244	3.164	-6.709	6.974

ΔNationalBankDistress are asset-weighted bank stock returns, winsorized at 1st and 99th percentile. See Table A.2.10 for all data sources. Source Ireland: IMF (2010); source Spain: New York Times (2012); source Portugal: Reuters (2014); source Italy: Financial Times (2016); source Greece: Reuters (2010).

2.3.2 Instrumenting Bank Distress using Exposure-Weighted Stock Market Returns

Our goal is to derive a variation in $NationalBankDistress_{it}$ that is, first, unaffected by omitted variables representing Eurozone-specific developments or risk attitudes. Secondly, the instrument should not increase simultaneously with national sovereign distress of the respective Eurozone country, thus limiting the biasing impact of reverse causality. In a nutshell, a valid instrument to identify the transmission of bank-to-sovereign distress in the Eurozone needs to have a strong correlation with national bank distress (relevance condition) and affect sovereign distress only through its impact on national bank distress while being uncorrelated with unobserved factors in the error term (exclusion restriction).

We argue that economic shocks that occur outside the Euro Area but directly impact the credit exposure and thereby the performance of EZ banks can fulfill these criteria, as discussed in the following.[9] We measure imported fragility from outside the EZ by focusing on stock returns of non-EZ countries where the respective EZ banking sector is invested in and weight these returns using the bilateral claims of the respective EZ banking sector.

Using the Consolidated Banking Statistics of the BIS, we collect the consolidated claims of a Eurozone country's banking sector i against all borrowers (banks, official sector, non-bank private sector) of a non-Eurozone country k. These claims are a suitable approximation for the international credit exposure of a country's banking sector as they aggregate all international claims, including those from banks' foreign affiliates, consolidated on the bank's headquarter level. In order to avoid banks' endogenous shifting of their international claims, for instance due to the financial or Euro crisis, we fix each weight at the first observation of 1999, i.e. the start of our sample period.[10] Also, we exclude the direct claims between Eurozone countries since they are subject to common Eurozone crisis factors and distress spillovers which is precisely the correlation we want to avoid.

We focus on the BIS's immediate counterparty basis claims as they cover the longest time window.[11] Since we require both the BIS exposure data towards and stock return series of a non-Eurozone country, we end up with 47 countries outside the EZ. The list of countries can be found in the Appendix (Table A.2.8) and covers the most important markets for Eurozone banks. Our data shows that the exposure towards these non-Eurozone countries, when converted to Euro, makes up on average 15.5% of the total asset size of a country's banking sector over our sample period, and is therefore meaningful enough to have an economic impact on its financial sector performance. Table 2.2 shows this ratio for all countries in the estimation.

---

[9]For example, a New York Times (2013) article from January 31, 2013 reports that Spanish bank Santander *"now generates half of its earnings in Latin America's emerging economies"* and that *"a slowdown in Brazil and Mexico, combined with financial troubles in Europe, weighed on Santander's earnings last year."* As a consequence *"shares in Santander fell 2.3 percent in morning trading in Madrid on Thursday after the bank's fourth-quarter earnings fell below analysts' expectations."*

[10]Data points before are of worse quality. For Portugal, Greece and Ireland, the first BIS claim weight is reported in 1999:Q4, 2003:Q4 and 2006:Q1 respectively. We fix the weights of these countries at these points in time.

[11]The BIS statistics show occasionally small gaps in the claim data. For statistics that use the full time period, we interpolate the data.

Table 2.2: Banking sector exposure towards non-Eurozone countries to total bank assets

Austria	Belgium	France	Greece	Ireland	Italy	Netherlands	Portugal	Spain
15.9	19.7	15.0	11.3	11.9	6.53	27.8	6.14	22.6

Ratio (in %) between a banking sector's BIS immediate counterparty claims towards non-Eurozone countries in Table A.2.8 and total bank assets (mean of 1999-2016 sample period).

However, one concern with the selection of non-Eurozone countries could be that some markets may depend heavily on credit supply from a Euro Area banking sector. As such countries might react more distinctly to Euro-related events, we define a criterion to be included in the final exposure portfolio: We list the claims of all the BIS's creditor countries (including those outside the Euro Area) towards a non-Eurozone country. We identify all creditors that hold the 90th percentile or more of the claims towards this non-Eurozone market. If a Eurozone country is at any point in our sample period a "large creditor" towards a non-Eurozone country based on this definition, this non-Eurozone country is removed from the Eurozone country's exposure portfolio.[12]

The claim of each EZ country i towards a non-EZ country k ($BankClaim_{ik}$) in 1999 is then set in relation to the total claims of i towards all K countries (excluding "large" creditor/borrower relationships) in the sample:

$$Weight_{ik} = \frac{BankClaim_{ik}}{\sum\limits_{k=1}^{K} BankClaim_i} \tag{2}$$

$Weight_{ik}$ is therefore a measure for the importance of a non-Eurozone country k in the portfolio of the banking sector in Euro country i at beginning of our sample period.

In order to measure the distress of the exposure, we then multiply this weighting factor with a daily-varying stock market return series of country k:

$$ExposureWeightedReturns_{ikt} = Weight_{ik} * StockMarketReturns_{kt} \tag{3}$$

We use broad stock market series that encompass all industries of the respective non-Eurozone country so that the exposure weights match the stock returns by covering all sectors of the economy. Finally, these exposure weighted returns are aggregated on the country level of each country i to construct a series of $NonEZStockReturns_{it}$.

---

[12]Table A.2.9 in the appendix lists the "large creditor/borrower" pairs. In a robustness test, we alternatively drop European countries not part of the Euro Area and find similar results.

A final step we carry out to construct our instrument concerns the issue that international stock returns could be partially driven by Eurozone events due to the significance of the currency union for the global financial system. To deal with this concern, we aim to remove the Eurozone-specific component in global stock returns. To this end, we orthogonalize $NonEZStockReturns_{it}$ towards $EurozoneBankReturns_t$, i.e. the bank return series on the Eurozone level we derived in Section 2.3.1 that approximates pure Eurozone bank distress.[13]

We have reason to believe that our measure of exposure-weighted bank returns from outside the Eurozone is a valid instrument to isolate bank-to-sovereign distress shocks. First of all, bank stock returns in open economies such as the Eurozone have a sizeable global component and should therefore likely be affected by exposure-weighted return series from other countries. Negative shocks transmitted from these non-EZ markets can hurt EZ banks, e.g. by loan losses, asset write-downs or currency losses. Secondly, because these shocks are imported from other countries not part of the Euro currency area, they are less likely to be affected by unobserved factors regarding Euro Area politics or a Euro country's national sovereign fragility. The exogeneity is further strengthened as the final instrument is purged from Eurozone-specific bank return variation, uses fixed claim data from the beginning of the sample period and is removed of all non-Eurozone countries that depend heavily on Eurozone bank funding.

2.3.3 Set of Dependent and Explanatory Variables

2.3.3.1 Dependent Variable

Similar to Gerlach et al. (2010) and Singh et al. (2016) sovereign distress is measured as the spread between the 10-year sovereign bond return of an EZ country and the 10-year sovereign bond return of Germany, multiplied with minus one:[14]

$$\Delta SovereignDistress_{it} = (-1)*(NationalSovereignBondReturn_{it} -$$
$$GermanSovereignBondReturn_t)$$

(4)

Higher values indicate higher levels of sovereign distress. Data is taken from Datastream. Figures 2.1 to 2.5 depict ΔNationalBankDistress and ΔSovereignDistress for the five GIIPS

---

[13]We conduct a different test specification to remove the Eurozone-specific variation in the robustness section and find similar results.

[14]A robustness check using yields instead of returns yields similar results.

countries over different 3- or 4-month periods which include several events that affected bank and sovereign distress in the respective country. All figures show that our measures for bank and sovereign distress are highly positively correlated and respond to key events during the Eurozone crisis.[15] We again prefer bond yields over CDS spreads since some CDS turned illiquid during the Euro Area crisis which would require to disregard e.g. Greece from the analysis. However, we also use CDS spreads of sovereigns as a dependent variable in the robustness section and find similar effects.

Figure 2.1: Bank distress (based on ΔNationalBankDistress) and sovereign distress (based on ΔSovereign Distress) in Italy from May 1st 2016 to August 1st 2016

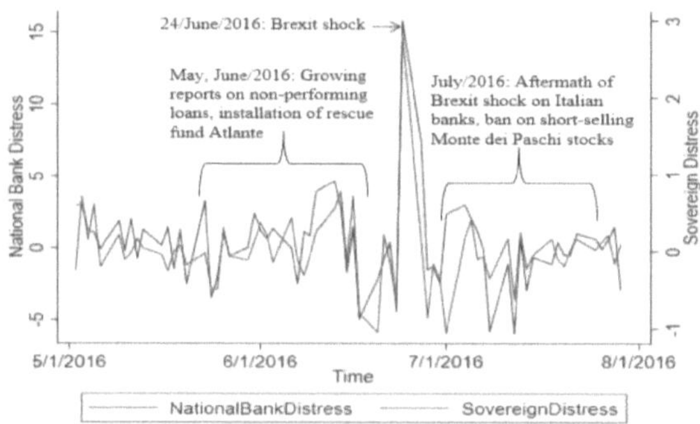

Figure 2.2: Bank and sovereign distress in Spain from April 1st 2012 to July 1st 2012

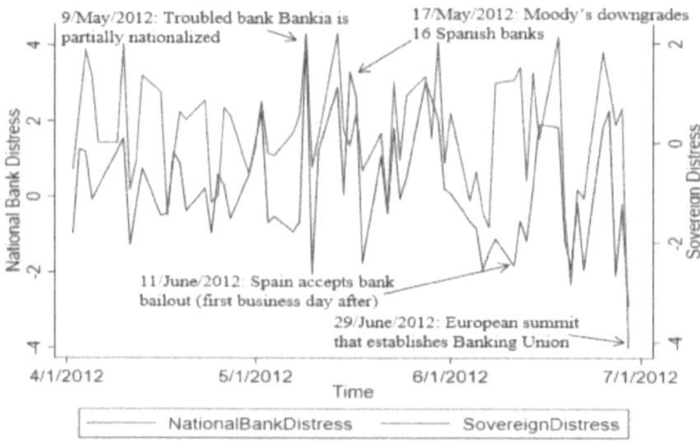

---

[15]Sources Italy: The Guardian (2016), CNN (2016); sources Spain: Financial Times (2012), Reuters (2012), New York Times (2012), The Guardian (2012); sources Portugal: The Guardian (2011a), The Guardian (2011b), Financial Times (2011); sources Ireland: The Guardian (2010b), New York Times (2010), The Guardian (2010e), Moody's (2010); sources Greece: CNBC (2010), The Guardian (2010a), The Guardian (2010d), Reuters (2010), The Guardian (2010c).

Figure 2.3: Bank and sovereign distress in Portugal from May 1st 2011 to August 1st 2011

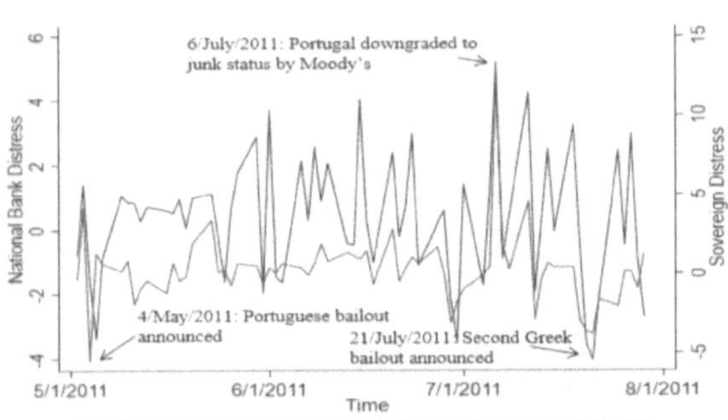

Figure 2.4: Bank and sovereign distress in Ireland from September 1st 2010 to January 1st 2011

Figure 2.5: Bank and sovereign distress in Greece from February 1st 2010 to May 1st 2010

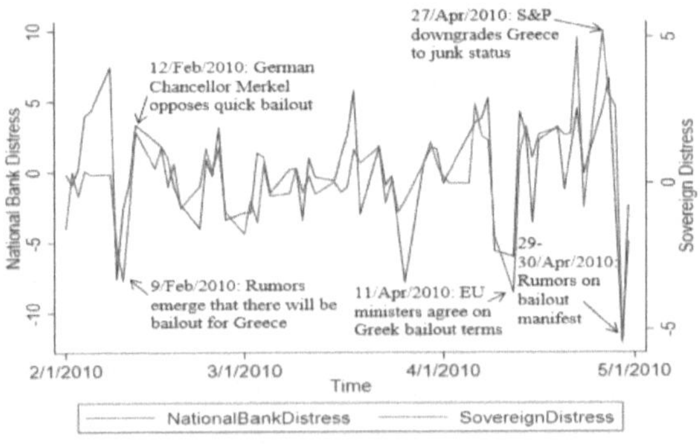

2.3.3.2 Control Variables

In order to control for the impact of daily developments in financial markets that could influence both sovereign and bank distress, we introduce a broad set of explanatory variables to capture international, European and national financial market developments. Definitions and sources of all control variables can be found in Table A.2.10.

Global factors have been shown to drive sovereign creditworthiness to a sizeable degree (Longstaff et al. 2011). We therefore control for the VIX to capture the implied volatility of US equity markets to capture volatility, "fear", or increased risk premiums in US financial markets. We also control for the US term spread, i.e. the yield spread of a 10-year US treasury bond and a 3-month T-Bill which approximates the premium investors receive for long-term investments.

On a European level, we control for changes in the VSTOXX which capture similar volatility dynamics than the VIX but are based on the EuroStoxx50. We expect increases in the VSTOXX to be associated with heightened financial market volatility and thus rising sovereign distress. Also, we incorporate the nominal effective exchange rate of the Euro into our analysis, i.e. the weighted exchange rate of the Euro against the Eurozone's most important trading partners. This variable represents movements in the external value of the Euro. A lower external value could indicate better export opportunities for firms in the Eurozone which could lead to lower sovereign distress. However, a depreciation of the Euro could also be the result of negative news shocks regarding the Eurozone crisis that are associated with higher sovereign spreads. We use the nominal effective exchange rate provided by JP Morgan but our framework is robust towards the version of other providers.

The ECB's monetary policy, especially its unconventional programs, is likely to impact both sovereign and bank distress. In order to account for these effects, we use the current account holdings of EZ banks at the ECB as a measure for the general stance of monetary policy. The current account holdings are the sum of banks' required and excess reserves held at the ECB and expanded considerably as a consequence of the central bank's asset purchase programs, i.e. SMP and PSPP. We believe expansionary monetary policy that is visible in the current account holdings to have a negative impact on sovereign spreads.

The term spread on a Eurozone-level might differ in its informativeness regarding short- and long-term interest rates compared to the US version. Hence, we also control for the spread between a 7-10 year FTSE MTS Eurozone government broad yield, which is designed to be a measure of the overall interest rate level in the Eurozone government bond market, and the 3-month Euribor rate.

Finally, we want to account for macroeconomic factors on the country level that go beyond the financial sector impact picked up by our bank distress measure. If, for instance, firm distress due to the recession in the Eurozone drives up sovereign spreads and impairs the banking sector through non-performing loans, not controlling for these effects could hamper our statistical inferences. We therefore include each EZ country's stock return index comprising non-financial firms in our panel, provided by Datastream, as a covariate. As we are interested in the non-financial-specific variation of these stock prices and not the co-movement with banking sector distress, we orthogonalize the non-financial stock returns towards the bank distress measure of every country. That way, which is similar to the approach of Beck et al. (2017), we are provided with a measure of stock return shocks specific to the real-sector of a country. In order to separate these shocks further from common Eurozone-wide effects, we, similar to our construction of country-specific bank distress, also orthogonalize the non-financial stock returns with respect to a total Eurozone stock return series provided by Datastream. The resulting returns are now both country- and non-financial-sector-specific. We test our model also for other versions of this variable in the robustness section and find concordant results.

All variables we use are winsorized at the 1st and 99th percentile to alleviate the impact of outliers. Subsequently, the variables are standardized to ease the interpretation of the estimated economic effects. As a rule of thumb, we use variables in natural logs that have no sensible meaning in the un-manipulated level, such as bank stock or sovereign bond prices or exchange rates. Variables in percentages such as spreads or interest rates have a natural meaning in levels therefore we do not use them in logs. Since most variables in levels and log levels are not stationary, we use all (logged) variables in first differences. Summary statistics of the non-standardized variables can be found in Table 2.3.

Table 2.3: Summary statistics of all variables used in the main empirical specification

Variable	Obs.	Mean	Median	Std. Dev.	Min	Max
ΔSovereignDistress	42,192	0.00191	0	0.393	-1.670	1.728
ΔNationalBankDistress	40,555	0.0442	-0.00306	2.020	-6.709	6.974
NonEZStockReturns	38,999	0.00723	0.0471	1.047	-3.443	3.034
ΔVstoxx	42,246	-0.0130	-0.0600	1.502	-4.480	5.130
ΔVIX	42,255	-0.00682	-0.0400	1.386	-4.170	5.030
ΔNominalExchangeRate	42,255	0.000518	0	0.369	-1.015	1.003
ΔCurrentAccountHoldings	42,255	-0.196	-0.100	9.835	-33.45	60.99
ΔUSTermSpread	42,255	0.000441	0	0.0603	-0.162	0.192
ΔEurozoneTermSpread	42,255	6.96e-05	0	0.0403	-0.110	0.122
NonFinancialStockReturns	40,561	0.00144	0.00511	0.919	-2.698	2.664

Sample period is 01/01/1999–31/12/2016. All variables are in daily growth rates, winsorized at the 1st and 99th percentile and depicted in non-standardized format. See Table A.2.10 for definitions and sources.

2.4 Empirical Specification and Results

To get a first indication of the effect of national bank distress on sovereign distress we estimate the following OLS regression for all nine Eurozone countries in our panel from $t = 01/01/1999 - 12/31/2016$:

$$\Delta SovereignDistress_{it} = \beta_1 \Delta NationalBankDistress_{it} + \beta_x \Delta Controls_{(i)t} + \alpha_i + \delta_t + \epsilon_{it} \quad (5)$$

$\Delta SovereignDistress_{it}$ measures the daily sovereign bond return of EZ country i minus Germany's sovereign bond return. $\Delta NationalBankDistress_{it}$ measures the country-specific banking sector returns. Both distress measures enter with converted signs so that higher values indicate higher levels of distress. $\Delta Controls_{(i)t}$ encompass all explanatory variables introduced in the previous section, specified either in simple first differences or natural log first differences. α_i are country fixed effects to address the possibility that both sovereign creditworthiness and financial sector distress are driven by time-invariant country-specific unobservable factors, such as customs and culture in the financial market structure and regulation. We also include time fixed effects δ_t on the quarter-year level to alleviate concerns that our results are influenced by time-specific market-wide developments that have a common effect on all countries. We cluster standard errors at the country level to allow for the correlation of unobserved factors in the error terms within countries. We expect a positive β_1, i.e. higher levels of banking sector fragility are associated with higher sovereign distress.

However, as highlighted above, a fundamental shortcoming of this OLS approach is that it produces possibly biased estimates of β_1 as it could be prone towards reverse causality between bank and sovereign distress and unable to control for unobserved crisis-related factors driving both distress sources. We use an IV regression to address these endogeneity concerns. The first and second stage of the IV approach are as follows:

$$\Delta NationalBankDistress_{it} = \gamma_1 NonEZStockReturns_{it} + \gamma_x \Delta Controls_{(i)t} + \alpha_i + \delta_t + \phi_{it} \quad (6)$$

$$\Delta SovereignDistress_{it} = \lambda_1 \Delta \widehat{NationalBank}Distress_{it} + \lambda_x \Delta Controls_{(i)t} + \alpha_i + \delta_t + \epsilon_{it} \quad (7)$$

Controls and fixed effects refer to the same variables used in the OLS framework. The instrument $NonEZStockReturns_{it}$ measures the exposure weighted stock returns of non-EZ countries, against which the respective Eurozone banking sector holds claims. We run OLS and IV over those points in time for which both approaches have overlapping data to allow a direct comparison. By drawing the predicted values from the first stage regression ($\widehat{NationalBank}Distress_{it}$), we explicitly exploit the variation in bank distress in the Eurozone that is due to international returns in stock markets that are tailored to the exposures of Euro Area banking sectors. Our approach tests two hypotheses: First, if bank distress instrumented by foreign exposure shocks affects Eurozone sovereign distress, which can be ascertained by the $NationalBankDistress$ coefficient of the second IV-stage. Second, if OLS and IV estimates of the sovereign-bank loop significantly differ, which can be detected by the difference in coefficient magnitudes of $NationalBankDistress$ of both approaches.

The results of the OLS and the IV estimation are reported in Table 2.4. All variables are standardized which eases the interpretation of the relative economic significance of the variables. The estimated outcomes of the first IV stage in column (2) suggest that our instrument of exposure-weighted stock returns in non-Eurozone countries is highly statistically significant in explaining national-specific banking sector distress of Eurozone countries. An increase of exposure-weighted non-EZ stock returns by one standard deviation is associated with a decrease in national-specific bank distress by 0.427 standard deviations on average. The F-statistic of 143.19 suggests that the instrument is unlikely to be weak. Comparing the standardized coefficients reveals that our instrument is by far the most influential determinant of national banking sector distress with a standardized coefficient of -0.427. The standardized coefficients of the VSTOXX (0.169) and VIX (-0.042) are much smaller (other determinants are statistically insignificant) underlining the economic importance of our instrument.

The second stage, reported in column (3), estimates the impact of the predicted values of national-specific bank distress on sovereign distress.[16] We find that the isolated transmission of imported bank distress to sovereign distress in the IV regression is both economically and statistically highly significant at the 1% level: An increase of instrumented national-specific banking sector distress by one standard deviation yields a 0.0828 standard deviation increase in sovereign distress. This result speaks in favor of the hypothesis that non-EZ exposure returns affect EZ bank distress, which then impacts sovereign distress. While the effect is sizeable, it is, however, substantially lower than the corresponding OLS coefficient of 0.167 reported in column (1). The instrumental variable coefficient is thus around 50% lower than the corresponding OLS estimate. Both coefficients are statistically significantly different at the 10% level.[17] These results, addressing our second hypothesis, suggest that estimating the sovereign-bank loop with simple OLS could lead to an over-estimation of the corresponding coefficients due to reverse causality or omitted variables.

Judging the relative economic importance of determinants of sovereign distress reveals that the standardized impact of the instrumented national banking sector distress (being 0.083) is lower than the standardized impact of the VSTOXX, the exchange rate, and the term spreads of the Eurozone and the US, while the standardized impact is higher than for the non-financial stock return, current account holdings, and the VIX.

We believe that this procedure is the most thorough way of isolating the bank-to-sovereign distress channel as our instrument takes only non-Eurozone shocks into account and these shocks are calculated with constant exposure weights, disregard Eurozone-dependent borrowers and are purged from Eurozone-specific variation. Our results indicate that concerns about reverse causality and potentially omitted variable bias in the sovereign-bank loop literature are likely valid as both IV and OLS coefficients differ by a significant order of magnitude. However, since the isolated bank-to-sovereign distress coefficient is statistically and economically significant, we provide evidence that banking sector distress was not just a by-product or a correlation, but a major factor and likely cause for deteriorating sovereign creditworthiness in the Euro Area and for instance during the Eurozone crisis.

---

[16]We estimate both stages in a single procedure to receive correct standard errors. However, in this way the predicted values from the first stage are not standardized when they enter the second stage and thus the coefficient size cannot be compared to the OLS version. As a solution, we estimate both IV stages separately and standardize the predicted values to receive comparable coefficient sizes, while extracting standard errors from the conventional 2sls manner. Also, note that the second stage of an IV typically does not report R^2.

[17]Chi-squared of 3.04; this test also corresponds to the Durbin-Wu-Hausman test.

Regarding the control variables, we find reasonable signs and significance levels on all specifications. One questionable sign could be the negative impact of the VIX on sovereign distress in both the IV and the OLS estimation. However, this effect only emerges once the VSTOXX is included as a covariate. This result indicates that the VIX picks up the remaining international market volatility. Significance levels suggest that controlling for market volatility in Europe through the VSTOXX is more important than international volatility in the VIX.

2.5 Robustness

2.5.1 Comparison of IV and OLS during Eurozone Crisis

While we run our main regression from 1999 onward to exploit maximum data availability, we focus in this subsection on the bank-to-sovereign distress transfer during the Euro crisis that started in 2009. To make the comparison between OLS and IV estimation as precise as possible, we winsorize, standardize and orthogonalize all our variables only for the post 2008 period. Furthermore, we use fixed BIS exposure from 2007:Q1, as they are likely more informative than 1999:Q1 weights we used so far.

Results are reported in Table 2.5. For the first stage of the IV estimated for the years 2009 to 2016, we obtain a coefficient of non-Eurozone stock returns of 0.395 with an F-value of 111.28 reported in column (1) of Table 2.5. The instrument is therefore still highly statistically significant albeit slightly smaller than for the full sample. For the second stage of the IV estimation, the standardized coefficient of bank distress is at 0.112 and statistically significant at the 1% (column (2)). This larger coefficient compared to the baseline is reasonable, given that bank-to-sovereign distress transfers were especially pronounced during the Euro crisis. The corresponding OLS coefficient is 0.199 and thus the relative difference in the size of the coefficients is comparable to our main specification (column (3)). However, given the shorter estimation period, this difference in magnitudes is more noteworthy, as IV and OLS coefficients are now statistically different from another at the 3% level. Hence, if anything, the importance of taking reverse causality effects into account matters more during the crisis period.

Table 2.4: Baseline results: transmission of bank-sovereign distress with OLS and IV

	OLS	IV-2SLS	
	(1) ΔSovereignDistress	(2) ΔNationalBankDistress	(3) ΔSovereignDistress
ΔNationalBankDistress	0.167**		0.0828***
	(0.0511)		(0.0175)
NonEZStockReturns		-0.427***	
		(0.0357)	
ΔVIX	-0.00968*	-0.0421*	-0.00807
	(0.00486)	(0.0185)	(0.00575)
ΔUSTermSpread	-0.0988***	-0.00920	-0.0990***
	(0.0244)	(0.00570)	(0.0232)
ΔVstoxx	0.151***	0.169**	0.154***
	(0.0325)	(0.0583)	(0.0352)
ΔNominalExchangeRate	-0.108***	0.0142	-0.108***
	(0.0239)	(0.0354)	(0.0233)
ΔEurozoneTermSpread	0.108***	-0.00945	0.108***
	(0.0154)	(0.00552)	(0.0142)
ΔCurrentAccountHoldings	-0.0156***	0.00257	-0.0156***
	(0.00343)	(0.00361)	(0.00327)
NonFinancialStockReturns	-0.0736*	0.0353	-0.0731**
	(0.0329)	(0.0773)	(0.0300)
Constant	-0.0264*	-0.0396	0.0317**
	(0.0136)	(0.0242)	(0.0133)
Observations	37,794	37,794	37,794
R-squared	0.112	0.273	
Number of Countries	9	9	9
Time & Country FE	Yes	Yes	Yes
Chi^2 between OLS and IV of ΔNationalBankDistress			3.04*

This table shows the effects of ΔNationalBankDistress on ΔSovereignDistress for 9 Eurozone countries (Austria, Belgium, France, Greece, Ireland, Italy, Netherlands, Portugal, Spain) during the period 01/01/1999 to 12/31/2016. ΔSovereignDistress is the daily change in the natural logarithm of a country's 10-year government bond index relative to Germany's respective bond index change. ΔNationalBankDistress are asset-weighted bank stock returns on the country-level minus asset-weighted bank stock returns on the Eurozone-level. Estimated coefficients in (1) are from least squares regression. Column (2) instruments ΔNationalBankDistress with weighted stock market returns from non-Eurozone countries that are weighted according to the BIS claims of the Eurozone country towards all borrowers in the respective non-Eurozone country (NonEZStockReturns). Column (3) shows the 2nd stage of this IV regression in which ΔNationalBankDistress refers to the predicted values from (2). ΔVIX is the daily change in the VIX volatility index, ΔUSTermSpread is the daily change between the 10-year US Treasury yield and the 3-month T-Bill yield, ΔVstoxx is the daily change in the Vstoxx volatility index, ΔNominalExchangeRate is the change in the natural logarithm of the nominal effective exchange rate of the Euro, ΔEurozoneTermSpread is the daily change in the spread between a Eurozone 7-10 year broad yield and the 3-month Euribor yield, ΔCurrentAccountHoldings is the daily change in the natural logarithm of current account holdings (i.e. minimum and excess reserves held by banks at the ECB) and ΔNonFinancialStockReturns are the daily changes in the natural logarithm of a country's non-financial stock market returns, which are orthogonalized towards ΔNationalBankDistress and total stock market returns in the Eurozone. Chi^2 shows the test statistic for the statistical difference between ΔNationalBankDistress in the OLS and the IV estimation. All variables are standardized and winsorized at the 1st and 99th percentile. All columns include country and time fixed effects on the quarterly level. Standard errors (in parentheses) are clustered at the country level, ***, ** and * indicate statistical significance at the 1%, 5% and 10% level, respectively. See Table A.2.10 for variable definitions and sources.

Table 2.5: Robustness: comparison of IV and OLS during Eurozone crisis

	OLS	IV-2SLS	
	(1) ΔSovereignDistres	(2) ΔNationalBankDistress	(3) ΔSovereignDistress
ΔNationalBankDistress	0.199*** (0.0510)		0.112*** (0.0318)
NonEZStockReturns		-0.395*** (0.0375)	
Observations	18,213	18,213	18,213
R-squared	0.175	0.275	
Number of Countries	9	9	9
Time & Country FE	Yes	Yes	Yes
Controls	Yes	Yes	Yes
Chi^2 between OLS and IV of ΔNationalBankDistress			4.77**

This table shows robustness checks for the effects of ΔNationalBankDistress on ΔSovereignDistress with respect to the estimation period, which ranges from 01/01/2009 to 12/31/2016 to capture the Euro crisis. In contrast to Table 2.4, this specification also winsorizes, orthogonalizes and standardizes all data from 2009 onward. Fixed BIS claim weights are from 2007:Q1. Estimated coefficients in (1) are from least squares regression. Column (2) instruments ΔNationalBankDistress with weighted stock market returns from non-Eurozone countries (NonEZStockReturns). Column (3) shows the 2nd stage of this IV regression. Chi^2 shows the test statistic for the statistical difference between ΔNationalBankDistress in the OLS and the IV estimation. All variables are standardized and winsorized at the 1st and 99th percentile. 9 Eurozone countries are Austria, Belgium, France, Greece, Ireland, Italy, Netherlands, Portugal and Spain. All columns include country and time fixed effects on the quarterly level and the daily control variables discussed in Table 2.4. Standard errors (in parentheses) are clustered at the country level, ***, ** and * indicate statistical significance at the 1%, 5% and 10% level, respectively. See Table A.2.10 for variable definitions and sources.

2.5.2 Alternative Versions of the Dependent Variable

We perform a range of sensitivity analyses to demonstrate the robustness of our results. First, we replace the sovereign bond return spreads with sovereign bond yield spreads. Also, we conduct the estimation using the 5-Year CDS premium of the respective country minus the 5-Year CDS premium of Germany (both quotes in US dollars) as the dependent variable to show that are our results do not depend on using government bonds as a sovereign fragility indicator. For the CDS version, we have to disregard Greece from the analysis, as its CDS rate turned illiquid during the course of the Euro crisis. Also, CDS data only starts around 2008 which shrinks our sample considerably. Although this limits the comparability to our baseline somewhat, we want to make sure that our main results are broadly intact when using CDS data. Columns (1) and (2) in Table 2.6 report the second stages of the IV-2SLS regressions with the results of the versions using sovereign bond yields and CDS spreads. We find similar results on both specifications compared to our baseline. Most control variables enter with the same sign and significance. More importantly, the coefficient of instrumented bank distress has the same level of statistical significance and a comparable size in all specifications, in particular for sovereign yields, though it is slightly larger when using CDS spreads.

Table 2.6: Robustness: alternative versions for dependent variable

	(1) ΔSovereign Yield Spread	(2) ΔCDS Spread
ΔNationalBankDistress	0.0724***	0.114***
	(0.0161)	(0.0192)
ΔVstoxx	0.144***	0.122**
	(0.0348)	(0.0479)
ΔNominalExchangeRate	-0.103***	-0.126***
	(0.0233)	(0.0324)
ΔEurozoneTermSpread	0.117***	0.109***
	(0.0169)	(0.0287)
ΔUSTermSpread	-0.0802***	-0.0816***
	(0.0204)	(0.0225)
ΔVIX	-0.00697	-0.0212***
	(0.00632)	(0.00756)
NonFinancialStockReturns	-0.0731**	-0.0848***
	(0.0313)	(0.0306)
ΔCurrentAccountHoldings	-0.0115***	0.000536
	(0.00311)	(0.00311)
Observations	37,586	17,210
Number of Countries	9	8
Time & Country FE	Yes	Yes

This table shows robustness checks for the effects of ΔNationalBankDistress on forms of ΔSovereignDistress with respect to the dependent variable in the main specification. Columns show the second stage of the IV-2SLS estimation in which country-specific bank distress of 9 Eurozone countries (Austria, Belgium, France, Greece, Ireland, Italy, Netherlands, Portugal and Spain) is instrumented using exposure-weighted non-Eurozone stock returns from 01/01/1999 to 12/31/2016 according to Table 2.4. Column (1) uses the spread in 10-year sovereign bond yields between a country and the German rate as a dependent variable. Column (2) repeats the analysis with the 5-year CDS rate of a country with respect to the German CDS rates. Due to data constraints, this specification excludes Greece and starts in 2008. All variables are standardized and winsorized at the 1st and 99th percentile. All columns include country and time fixed effects on the quarterly level and the daily control variables discussed in Table 2.4. Standard errors (in parentheses) are clustered at the country level, ***, ** and * indicate statistical significance at the 1%, 5% and 10% level, respectively. See Table A.2.10 for variable definitions and sources.

2.5.3 Alternative Versions for Bank Distress Variable

Next, we check if the effects we derived were due to our definition of national-specific bank distress. We therefore repeat the baseline analysis using the weighted bank stock returns on the country level but without subtracting any Eurozone-specific component ($BankReturns_{it}$, here multiplied times minus one).[18] Column (1) in Table 2.7 shows that the instrument for this estimation is also strong and highly significant. Results in column (2) yield a highly

---

[18]To estimate in accordance with our baseline approach, we now orthogonalize the non-financial returns on the county level with respect to the EZ stock returns and $BankReturns_{it}$, as this is now the right-hand-side variable of interest.

statistically significant effect of instrumented bank distress on the second stage of the IV. The coefficient is somewhat larger than in the baseline which could undermine our conjecture that the original OLS version seems to overestimate the transmission of distress in the sovereign-bank loop. Therefore, column (3) reports the OLS estimation using the same unadjusted bank distress on the country level as a right-hand side measure of interest. In this case, the respective coefficient is also larger in size and surpasses the IV coefficient by almost 50%. This finding suggests that we can maintain our presumption that our IV specification accounts for the reverse causality and omitted variable biases present in the OLS estimation.

One further concern could be the way in which we removed the Eurozone-specific variation from the bank distress variable. We chose to subtract the Eurozone return index to follow as close as possible to Buch & Neugebauer (2011). However, one could also eliminate the EZ component by means of orthogonalization. Columns (4)-(6) in Table 2.7 show the two IV stages and the OLS estimation when following this approach. The results are extremely close to our baseline and the IV and OLS coefficients differ to a very similar degree.

2.5.4 Alternative Versions for Instrumental Variable

The next concern we address is the orthogonalization of the instrument by which we removed the Eurozone-specific variation in Section 2.3.2. We used the Eurozone return index we derived beforehand and orthogonalized it towards a world stock return series that excludes the EZ. As an alternative, we also utilize a previously derived variable, namely the total (i.e. not country-specific) bank returns of Eurozone countries constructed in Section 2.3.1 ($BankReturns_{it}$). We conduct a principal component analysis based on these bank returns of all EZ countries. The first component, approximating return variation that is common for all Eurozone countries, explains roughly 59% of the total variation. In order to isolate the Eurozone-specific part of the variation, we once again orthogonalize this first principal component with respect to the same world bank stock return series that excludes the EZ. Finally, we clean the instrument of this Eurozone-specific component by means of orthogonalization. The results of this different elimination procedure for both IV stages are reported in columns (1) and (2) of Table 2.8 in which we find almost identical results compared to our baseline.

Another potential concern related to our IV specification could be the removal of credit-dependent non-Eurozone countries in the construction of the instrument. We removed all non-

Table 2.7: Robustness: alternative versions for national bank distress variable

	IV-2SLS		OLS	IV-2SLS		OLS
	(1)	(2)	(3)	(4)	(5)	(6)
	ΔBankDistress	ΔSovereignDistress	ΔSovereignDistress	ΔNationalBankDistress by orthogonalization	ΔSovereignDistress	ΔSovereignDistress
NonEZStockReturns	-0.226***			-0.459***		
	(0.0323)			(0.0367)		
ΔBankDistress		0.152***	0.235***			
		(0.0325)	(0.0598)			
ΔNationalBankDistress by orthogonalization					0.0759***	0.151**
					(0.0161)	(0.0499)
Observations	37,794	37,794	37,794	37,794	37,794	37,794
R-squared	0.304		0.127	0.266		0.109
Number of Countries	9	9	9	9	9	9
Time & Country FE	Yes	Yes	Yes	Yes	Yes	Yes
Controls	Yes	Yes	Yes	Yes	Yes	Yes

This table shows robustness checks for the effects of ΔNationalBankDistress on ΔSovereignDistress with respect to changes in the specification of national bank distress. In the IV specifications, bank distress of 9 Eurozone countries (Austria, Belgium, France, Greece, Ireland, Italy, Netherlands, Portugal and Spain) is instrumented using exposure-weighted non-Eurozone stock returns from 01/01/1999 to 12/31/2016 according to Table 2.4. Column (1) estimates the first IV stage using the more broadly specified bank distress on the country level, i.e. without removing a Eurozone-specific component ($BankReturns_{it}$, here multiplied times minus one). This bank distress variable is instrumented using the same exposure-weighted foreign stock returns as before. Column (2) shows the result of the second IV stage using this variable and column (3) uses the same specification but applying OLS. Column (4) estimates the first IV stage using the national bank distress measure in which the Eurozone-specific component was not removed by subtraction but by orthogonalization. Columns (5) and (6) show the results of the corresponding second IV stage and OLS. All variables are standardized and winsorized at the 1st and 99th percentile. All columns include country and time fixed effects on the quarterly level and the daily control variables discussed in Table 2.4. Standard errors (in parentheses) are clustered at the country level, ***, ** and * indicate statistical significance at the 1%, 5% and 10% level, respectively. See Table A.2.10 for variable definitions and sources.

Eurozone countries in the exposure portfolio of a Eurozone country if this Eurozone country held at any point in our sample period more than 90th percentile of the total international claims towards the respective non-Eurozone country. Another way to remove countries that are economically close to the Euro Area and could face Eurozone crisis spillovers is to simply remove all EU countries which do not have the Euro as their main currency from the sample as they share similar institutions and regulation with the rest of the Eurozone (though several of these countries are also affected when removing the credit-dependent nations). We do so in columns (3) and (4) of Table 2.8, while otherwise specifying the instrument in the same way as before. Though the instrument is slightly weaker in the first stage, our main results are unaffected by this exercise.

Lastly, we consider the reverse case to the "large borrower" criterion we set up. Instead of removing non-EZ countries from the portfolio of an EZ country if the EZ country's claims against this non-EZ country exceed the 90th percentile of claims for this non-EZ market, we now disregard a non-EZ country if it is a "large" credit supplier to the EZ country, i.e. if the bilateral BIS liabilities of the EZ member vis-a-vis the respective non-EZ member is at the 90th percentile or above.

Such cases cover, for instance, countries that hold a lot of Eurozone sovereign debt, which could have impacted certain banking sectors during the Euro crisis. Moreover, shocks in the non-EZ country could have a direct price effect on EZ sovereign bonds due to large sovereign bond holdings, which are not transmitted by the EZ banking sector. In practice, this criterion affects mainly the US and the UK, but also Japan. One drawback of this specification is that the BIS reports claims towards Eurozone countries only for 14 members from our sample of non-EZ countries. Despite this limited comparability, we report (columns (5) and (6) in Table 2.8) a rather similar second stage of our IV specification when applying this criterion.

Table 2.8: Robustness: alternative versions for instrumental variable

	(1) ΔNationalBank Distress	(2) ΔSovereign Distress	(3) ΔNationalBank Distress	(4) ΔSovereign Distress	(5) ΔNationalBank Distress	(6) ΔSovereign Distress
NonEZStockReturns: Different elimination of EZ component	-0.409*** (0.0355)					
NonEZStockReturns: Excluding EU instead of dependent borrowers			-0.410*** (0.0440)			
NonEZStockReturns: Excluding large credit suppliers					-0.361*** (0.0472)	
ΔNationalBankDistress		0.0822*** (0.0175)		0.0868*** (0.0126)		0.0756*** (0.0166)
Observations	37,794	37,794	37,794	37,794	37,794	37,794
R-squared	0.266		0.266		0.272	
Number of Countries	9	9	9	9	9	9
Time & Country FE	Yes	Yes	Yes	Yes	Yes	Yes
Controls	Yes	Yes	Yes	Yes	Yes	Yes

This table shows robustness checks for the effects of ΔNationalBankDistress on ΔSovereignDistress with respect to changes in the specification of the instrumental variable. In the IV specifications, bank distress of 9 Eurozone countries (Austria, Belgium, France, Greece, Ireland, Italy, Netherlands, Portugal and Spain) is instrumented using exposure-weighted non-Eurozone stock returns from 01/01/1999 to 12/31/2016 according to Table 2.4. Column (1) estimates the first IV stage using a different orthogonalization procedure to remove the EZ component from the instrument. We conduct a principal component analysis of the asset-weighted bank stock returns of each Eurozone-country ($BankReturns_{it}$), as derived in Section 2.3.1. We orthogonalize the first component of these returns towards the world bank return factor. This variable gives us a Eurozone-specific bank stock variation and we purge our instrument from these factors by means of orthogonalization. Column (2) shows the second stage of the corresponding IV. Columns (3) and (4) estimate a specification in which we remove all EU countries in the construction of the instrument instead of the removal of non-EZ countries whose founding depends heavily on a Euro member. Columns (5) and (6) show the results when removing non-EZ countries that supply large amounts of funding to an EZ country. All variables are standardized and winsorized at the 1st and 99th percentile. All columns include country and time fixed effects on the quarterly level and the daily control variables discussed in Table 2.4. Standard errors (in parentheses) are clustered at the country level, ***, ** and * indicate statistical significance at the 1%, 5% and 10% level, respectively. See Table A.2.10 for variable definitions and sources.

2.5.5 Strengthening the Exclusion Restriction of the Instrument

One key assumption of the instrumental variable approach is that the instrument, in our case exposure-weighted stock market returns from non-EZ countries, affects the dependent variable, sovereign creditworthiness, only through the instrumented variable, i.e. an EZ country's banking sector distress. A potential concern for our identification strategy could be that non-EZ stock return shocks are transmitted to EZ sovereign distress through other channels than banking sectors. For instance, real economic downturns in a non-EZ country that are visible in falling stock returns could spill over to firms in a country of the Eurozone and worsen its sovereign creditworthiness, independent of banking sector claims. This effect could be more pronounced if the countries share stronger trading relationships. In this case, the exclusion restriction of the instrument would be violated.

Though we cannot categorically reject this channel, we have reason to believe that our approach is robust to these concerns. First, we already control for non-financial stock market returns of every EZ country which should account for real economic shocks transmitted to or stemming from any EZ economy. In addition, we also control for potential trade shocks affecting the Euro Area by including the nominal effective exchange rate of the Euro. We thereby account for non-financial shocks transmitted to sovereigns that could otherwise bypass the banking sector in our instrumental variable approach. Second, several papers show that trade or trade openness is not statistically significantly related to sovereign distress of Euro Area countries (Aizenman et al. (2013), Beirne & Fratzscher (2013)), suggesting that such shocks or any remaining biasing effect in our instrument would be limited in size.

We conduct further robustness tests to strengthen the exclusion restriction of the instrument. First, we build a version of exposure-weighted non-EZ stock market returns that consists only of bank stocks. By focusing on bank-specific stock market shocks of a non-EZ country as an instrument, it is less likely that this approach transmits distress to sovereign creditworthiness through non-financial sectors or trade-specific channels that could otherwise be present in the total stock market returns of a non-EZ country. Simply put, banking sector shocks of a non-Eurozone country are more likely to affect EZ sovereign creditworthiness in no other ways than through the banking sector of a Eurozone country. We construct this alternative specification in the same way as our baseline. Results in Table 2.9 column (1) show that the instrument is similar to the baseline in the first stage. Column (2) reports a slightly larger and still highly statistically significant second stage coefficient of instrumented bank distress that is close to our baseline. This result suggests that alternative channels of how non-EZ stock market shocks could affect sovereign creditworthiness in other ways than bank distress are, if present, limited in size and not critical for our results.

In a further robustness check, we aim to control directly for trade-related shocks. To do so, we construct an export-weighted non-financial stock market variable: export volumes of EZ to non-EZ countries are drawn from the IMF's Direction of Trade Statistics. We construct export weights that approximate the importance of a trading partner country in the same way as the BIS weights in Section 2.3.2. We multiply the export importance of a non-EZ country towards an EZ country with a non-financial stock market return series of the former since this variable is closer to capture real-economic variation that affects the trading performance of a country. Since we are interested in the trade-specific variation of this variable and not the co-movement with stock returns captured in the instrument, we orthogonalize the trade shock measure towards our instrumental variable. The trade shock variable enters our baseline as an additional control variable. Results in column (3) of Table 2.9 show that trade-specific stock market returns enter statistically significantly in the first stage of the instrument, indicating that trade-specific shocks drive part of the variation in EZ bank distress. However, while the instrument remains statistically significant in the first stage, column (4) shows that the second stage coefficient of instrumented bank distress is nearly identical to our baseline while the export-weighted returns are statistically indistinguishable from zero. There could be additional channels, such as trade credit shocks which are not conducted via the banking system, but could impact real-sector performance of a Eurozone country and hence sovereign distress. Though we cannot categorically rule out such further trade-related channels, we conclude that with our control variables and conducted tests, we have reason to believe that these channels are, if present, not a significant threat to our main results.

Table 2.9: Robustness: strengthening exclusion restriction

	(1) ΔNationalBank Distress	(2) ΔSovereign Distress	(3) ΔNationalBank Distress	(4) ΔSovereign Distress
NonEZBankStockReturns	-0.414*** (0.0404)			
ΔNationalBankDistress		0.110*** (0.0239)		0.0836*** (0.0150)
NonEZStockReturns			-0.466*** (0.0359)	
Trade-Weighted Shocks			-0.151*** (0.0328)	0.00475 (0.0178)
Observations	37,794	37,794	37,794	37,794
R-squared	0.287		0.289	
Number of Countries	9		9	
Time & Country FE	Yes	Yes	Yes	Yes
Controls	Yes	Yes	Yes	Yes

This table shows robustness checks for the effects of ΔNationalBankDistress on ΔSovereignDistress to strengthen the exclusion restriction of the IV approach. The columns show the first and second stages of the IV-2SLS estimation in which bank distress of 9 Eurozone countries (Austria, Belgium, France, Greece, Ireland, Italy, Netherlands, Portugal and Spain) is instrumented using exposure-weighted non-Eurozone stock returns from 01/01/1999 to 12/31/2016. Columns (1) and (2) show first and second stages when using only bank stocks in the exposure-weighted stock returns as an instrument. Columns (3) and (4) show first and second stages when repeating the baseline estimation of Table 2.4 but adding trade-weighted shocks as an additional control, i.e. export-weighted non-financial stock returns of non-EZ countries. All variables are standardized and winsorized at the 1st and 99th percentile. All columns include country and time fixed effects on the quarterly level and the daily control variables discussed in Table 2.4. Standard errors (in parentheses) are clustered at the country level, ***, ** and * indicate statistical significance at the 1%, 5% and 10% level, respectively. See Table A.2.10 for variable definitions and sources.

2.5.6 Weekly Frequency

The next concern we address is related to our data frequency. We use daily data in order to draw from a larger set of observations. However, daily data may be noisy. Even though we already winsorized our data to account for this possibility, we collapse the data to a weekly frequency and re-do the baseline OLS and IV estimations. We find a very similar second stage IV coefficient in size and statistical significance, while the OLS coefficient increases in size, thus widening the gap between OLS and IV (Table 2.10). Our main results are therefore robust to a lower data frequency.

Table 2.10: Robustness: weekly frequency

	OLS	IV-2SLS	
	(1) ΔSovereignDistress	(2) ΔNationalBankDistress	(3) ΔSovereignDistress
ΔNationalBankDistress	0.245** (0.0793)		0.0909*** (0.0283)
NonEZStockReturns		-0.533*** (0.0460)	
Observations	7,761	7,761	7,761
R-squared	0.189	0.363	
Time & Country FE	Yes	Yes	Yes
Number of Countries	9	9	9
Controls	Yes	Yes	Yes

This table shows robustness checks for the effects of ΔNationalBankDistress on ΔSovereignDistress with respect to a change in frequency of the data. The columns show the first and second stages of the IV-2SLS estimation in which bank distress of 9 Eurozone countries (Austria, Belgium, France, Greece, Ireland, Italy, Netherlands, Portugal and Spain) is instrumented using exposure-weighted non-Eurozone stock returns from 01/01/1999 to 12/31/2016. Column (1) repeats the estimation from Table 2.4 when collapsing the data to the weekly frequency and shows results for an OLS estimation. Columns (2) and (3) show results for IV-2SLS estimations. All variables are standardized and winsorized at the 1st and 99th percentile. All columns include country and time fixed effects on the quarterly level and the weekly control variables discussed in Table 2.4. Standard errors (in parentheses) are clustered at the country level, ***, ** and * indicate statistical significance at the 1%, 5% and 10% level, respectively. See Table A.2.10 for variable definitions and sources.

2.5.7 Alternative Control Variables and Time Fixed Effects

We want to make sure that certain daily control variables in the IV regression do not critically drive our results. One potential candidate to do so could be the non-financial stock market returns on the country level. In our baseline, we orthogonalized this variable with respect to the national-specific bank returns and the stock returns on the Eurozone level. However, the former adjustment renders the non-financial returns insignificant in explaining bank return variation in the first stage of the instrument. To investigate if this step has any consequences for our main results, we introduce a version of non-financial stock returns that is only orthogonal towards stock returns on the Eurozone level. Columns (1) and (2) in Table 2.12 report the first and second stage of the IV regression. The new non-financial return variable is now negative and significant in the first stage regression, i.e. higher non-financial stock returns are associated with lower bank distress. However, the magnitude of the second stage coefficient for predicted country-specific bank distress remains almost unchanged. This result suggests that our chosen specification of non-financial returns is not critical for our main findings.

Another control variable that might require an additional robustness check is the VIX as it could correlate excessively with the VSTOXX. However, when we remove the VIX from the list of covariates, our results hardly change (column (3) of Table 2.12). Furthermore, when conducting the variance inflation factor (vif) test for our first and second IV stage, we find no covariate near the critical value of 10 (highest value 3.51, see Table 2.11), implying that multicollinearity in the controls is not a concern. In a further test, we replace the nominal Euro exchange rate from JP Morgan with the nominal Euro/US Dollar exchange rate, finding again similar effects (column (4)). Also, we replace quarterly with monthly time fixed effects in our IV regression to account for market-wide changes on a higher frequency. Column (5) indicates that our results are not sensitive to this adjustment. Finally, we add a trading liquidity variable to control for potentially low turnovers of certain bank stocks. We proceed in a similar way than for our bank distress measure and calculate the ratio of the daily turnover of a bank stock over the stocks outstanding for this bank. We then weight this ratio with a bank's asset size on the country level and add up the asset-weighted liquidity ratios on the country level. This variable is introduced as an additional control in our main regression (column (6)). The liquidity indicator is statistically insignificant and leaves our main results unchanged.

Table 2.11: Variance inflation factor (vif) test for estimations of Table 2.4

Variable	VIF (OLS)	VIF (1st stage IV)	VIF (2nd stage IV)
NonEZStockReturns		2.30	
ΔNationalBankDistress	1.24		3.51
ΔVIX	1.48	1.99	1.79
ΔUSTermSpread	1.10	1.10	1.10
ΔVstoxx	1.62	1.77	2.77
ΔNominalExchangeRate	1.19	1.21	1.20
ΔEurozoneTermSpread	1.05	1.05	1.05
ΔCurrentAccountHoldings	1.01	1.01	1.01
ΔNonFinancialStockReturns	1.21	1.21	1.24

Table 2.12: Robustness: changing control variables and time fixed effects

	(1) ΔNationalBank Distress	(2) ΔSovereign Distress	(3) ΔSovereign Distress	(4) ΔSovereign Distress	(5) ΔSovereign Distress	(6) ΔSovereign Distress
ΔNationalBankDistress		0.0845***	0.0767***	0.0662***	0.0800***	0.0833***
		(0.0182)	(0.0158)	(0.0165)	(0.0164)	(0.0176)
NonEZStockReturns	-0.427***					
	(0.0357)					
NonFinancialStockReturns			-0.0728**	-0.0662**	-0.0731**	-0.0734**
			(0.0308)	(0.0302)	(0.0295)	(0.0302)
NonFinancialStockReturns: Alt. Version	-0.288***	-0.0663				
	(0.0760)	(0.0404)				
ΔVIX	-0.0407*	-0.00928*		-0.0130**	-0.00804	-0.00814
	(0.0186)	(0.00552)		(0.00639)	(0.00585)	(0.00569)
ΔEuroDollar ExchangeRate				0.107***		
				(0.0218)		
ΔTradingLiquidity						-0.00689
						(0.0105)
ΔUSTermSpread	-0.0142**	-0.0990***	-0.0986***	-0.100***	-0.0988***	-0.0988***
	(0.00573)	(0.0234)	(0.0233)	(0.0233)	(0.0233)	(0.0233)
ΔVstoxx	0.116**	0.160***	0.155***	0.158***	0.154***	0.154***
	(0.0458)	(0.0355)	(0.0349)	(0.0355)	(0.0356)	(0.0352)
ΔNominalExchangeRate	-0.109***	-0.108***	-0.109***		-0.106***	-0.108***
	(0.0303)	(0.0274)	(0.0234)		(0.0224)	(0.0234)
ΔEurozoneTermSpread	-0.00221	0.108***	0.108***	0.110***	0.107***	0.108***
	(0.00409)	(0.0143)	(0.0143)	(0.0145)	(0.0141)	(0.0142)
ΔCurrentAccountHoldings	0.00336	-0.0157***	-0.0156***	-0.0141***	-0.0159***	-0.0156***
	(0.00381)	(0.00328)	(0.00325)	(0.00301)	(0.00336)	(0.00327)
Observations	37,794	37,794	37,794	37,794	37,794	37,794
R-squared	0.335					
Number of Countries	9	9	9	9	9	9
Quarterly Time & Country FE	Yes	Yes	Yes	Yes	No	Yes
Monthly Time & Country FE	No	No	No	No	Yes	No

This table shows robustness checks for the effects of ΔNationalBankDistress on ΔSovereignDistress with respect to changes in the specification of certain control variables. The columns show either the first or second stage of the IV-2SLS estimation in which bank distress of 9 Eurozone countries (Austria, Belgium, France, Greece, Ireland, Italy, Netherlands, Portugal and Spain) is instrumented using exposure-weighted non-Eurozone stock returns from 01/01/1999 to 12/31/2016. ΔNationalBankDistress refers to the predicted values of country-specific bank distress from the respective specification. Columns (1) shows the first and column (2) the second IV-stage of the baseline regression in which non-financial stock returns of a country, which are originally orthogonalized to ΔNationalBankDistress and a stock return index for the Eurozone, are now only orthogonalized towards the Eurozone stock return index. The first stage is shown to see that this version of non-financial returns has a negative impact on bank distress. Column (3) removes the VIX when estimating the IV-baseline regression. Column (4) uses the change in the natural logarithm of the Euro-to-Dollar exchange rate instead of the Euro's nominal effective exchange rate. Column (5) conducts the IV-baseline regression using monthly instead of quarterly time fixed effects. Column (6) adds a liquidity variable for the turnover of bank stocks. All variables are standardized and winsorized at the 1st and 99th percentile. All columns include country and time fixed effects on the quarterly (or monthly) level and the daily control variables discussed in Table 2.4. Standard errors (in parentheses) are clustered at the country level, ***, ** and * indicate statistical significance at the 1%, 5% and 10% level, respectively. See Table A.2.10 for variable definitions and sources.

2.5.8 Wild Cluster Bootstrapping

Lastly, we take concerns by Cameron et al. (2008) on downward biased standard errors if the number of clusters is relatively low into account. Using the wild bootstrap developed by Roodman et al. (2019), we find and report almost identical wild cluster bootstrapped t-values for both our first and second stage of our baseline IV in Columns (2) and (4) of Table 2.13.

Table 2.13: Robustness: bootstrapping standard errors

	(1) ΔNationalBank Distress	(2) ΔNationalBank Distress	(3) ΔSovereign Distress	(4) ΔSovereign Distress
NonEZStockReturns Baseline	-0.427*** [−11.97]			
NonEZStockReturns Wild Bootstrapping		-0.427*** [−11.965]		
ΔNationalBankDistress Baseline			0.0828*** [4.73]	
ΔNationalBankDistress Wild Bootstrapping				0.0828*** [4.71]
Observations	37,794	37,794	37,794	37,794
R-squared	0.273	0.273		
Number of Countries	9	9	9	9
Time & Country FE	Yes	Yes	Yes	Yes
Controls	Yes	Yes	Yes	Yes

This table shows robustness checks for the effects of ΔNationalBankDistress on ΔSovereignDistress in which standard errors are bootstrapped. The columns show either the first or second stage of the IV-2SLS estimation in which bank distress of 9 Eurozone countries (Austria, Belgium, France, Greece, Ireland, Italy, Netherlands, Portugal and Spain) is instrumented using exposure-weighted non-Eurozone stock returns from 01/01/1999 to 12/31/2016. Column (1) shows results of the first and column (3) results of the second IV stage as estimated in the benchmark model in Table 2.4. Columns (2) and (4) repeat the estimation but are adjusted by the wild cluster bootstrap method using Rademacher weights and setting replications to 1000. The resulting t-values are reported in square brackets. All variables are standardized and winsorized at the 1st and 99th percentile. All columns include country and time fixed effects on the quarterly level and the daily control variables discussed in Table 2.4. T-values (in square brackets) are clustered at the country level or bootstrapped, ***, ** and * indicate statistical significance at the 1%, 5% and 10% level, respectively. See Table A.2.10 for variable definitions and sources.

2.6 Conclusion

We present a novel instrumental variable approach to account for reverse causality and omitted variable biases in the estimation of the bank-to-sovereign distress transmission in the Eurozone. Banking sector distress of Eurozone countries is measured by the asset-weighted stock returns of 132 Eurozone banks. We instrument country-specific bank distress using the weighted stock returns of 47 non-Eurozone countries that are tailored to the BIS's banking sector claims of a Eurozone country towards borrowers in the respective non-Eurozone country. These imported shocks, aligned to the international exposures of Eurozone banking sectors, are shown to be a highly significant instrument for bank distress in the Eurozone. We carry out several adjustments in order to avoid reverse causality and omitted variable issues: First, we use fixed exposure weights from the beginning of our sample. Second, we only include those non-Euro countries into a weighted stock portfolio that do not depend heavily on credit supply by the respective Euro member country. Third, we purge the instrument from any Eurozone-specific variation by means of orthogonalization.

Controlling for a range of financial market indicators, we find a statistically and economically significant effect of instrumented banking sector distress on sovereign distress for the considered nine Eurozone countries in the period 1999-2016. Banking sector distress was therefore a major factor and likely cause for deteriorating sovereign creditworthiness during the crisis and not just a by-product or a correlation. The size of our instrumental variable coefficient is around 50% lower than the corresponding OLS coefficient, and the difference is statistically significant. This finding supports our conjecture of reverse causality and omitted variables in the sovereign-bank loop estimation which are uncontrolled for in the OLS framework. The statistical significance of the IV estimator and sizeable difference between OLS and IV coefficient holds for a large number of robustness checks.

Our results have straightforward ramifications for the debate on the future of the Eurozone. We showed that bank distress is a major determinant of the creditworthiness of sovereigns. This finding calls for the stringent participation of equity holders and junior creditors in the loss participation of bank bankruptcies which are currently governed by the bank recovery and resolution directive (BRRD). Applying these bail-ins predictably and credibly, while limiting exceptions for large or politically connected banks, could have the potential to lower this transmission of financial distress.

A.2 Appendix to Chapter 2

A.2.1 Drivers of Bank-to-Sovereign Distress Transmissions

This appendix chapter shows estimates of the drivers of the bank-to-sovereign risk transmission. It does not appear in the main text, as it was not part of the published version of the corresponding paper. The following specification uses the same data as the corresponding main section, however, it focuses on the Eurozone crisis from 2009 to 2016. All adjustments to the data are described in the first paragraph.

To investigate the potential drivers of bank-to-sovereign distress spillovers, we start with the IV-specification that we derived in Section 2.3.2. However, in order to focus on the Eurozone crisis, all estimations are from 01/01/2009 to 12/31/2016. Therefore, BIS claims used in the instrument are fixed towards the period of 2007:Q1, as this date is likely more informative than the 1999:Q1 weights used so far. Furthermore, we winsorize, standardize and orthogonalize all our variables with respect to the 2009-2016 period, similar to the robustness check conducted in Section 2.5.1. As they have full data coverage for the 2009-2016 period, the ΔUSCorporateSpread (daily change in the spread between the US corporate benchmark BBB 10-year yield and the respective AAA yield) by Thomson Reuters and the ΔNonFinancialItraxx (the residuals from a regression of the daily change in the natural logarithm of the 10-year Itraxx Europe against the corresponding change of the 10-year Itraxx senior and subordinated financial indices) by Markit are introduced as additional controls.

Table A.2.1 repeats the regressions of the main section in Table 2.4 that compare the size difference of bank-sovereign distress transmissions between OLS and IV estimates, but uses the adjusted 2009-2016 data. Both the OLS (0.195 instead of 0.167) and the IV coefficient (0.109 instead of 0.0828) for bank distress are slightly larger in the new specification, and extremely close to the robustness check in Section 2.5.1. This increase is likely because the sovereign-bank loop was most amplified during the 2009-2016 period. Nevertheless, the larger magnitude of the OLS towards the IV bank coefficient (80% in the new version compared to 100% in the main section) is very comparable in both specifications. In addition, OLS and IV estimates continue to differ statistically significantly. Therefore, the following specification is extremely close to the main part and thus appropriate to investigate the drivers of bank-to-sovereign risk transmissions during the Euro crisis.

Table A.2.1: Transmission of bank-sovereign distress: OLS and IV with 2009-2016 data

	OLS	IV-2SLS	
	(1) ΔSovereignDistress	(2) ΔNationalBankDistress	(3) ΔSovereignDistress
ΔNationalBankDistress	0.195***		0.109***
	(0.0502)		(0.022)
NonEZStockReturns		-0.427***	
		(0.0513)	
ΔVIX	-0.0280***	0.0274	-0.0385***
	(0.00414)	(0.0166)	(0.00491)
ΔUSCorporateSpread	0.0184*	0.0112*	0.0163*
	(0.00924)	(0.00586)	(0.00843)
ΔUSTermSpread	-0.106***	-0.0298***	-0.101***
	(0.0217)	(0.00740)	(0.0201)
ΔVstoxx	0.152***	0.0986***	0.123***
	(0.0364)	(0.0250)	(0.0336)
ΔNominalExchangeRate	-0.129***	0.0532	-0.124***
	(0.0264)	(0.0454)	(0.0224)
ΔEurozoneTermSpread	0.125***	0.0163**	0.122***
	(0.0174)	(0.00660)	(0.0162)
ΔNonFinancialItraxx	0.0194**	0.0318**	0.0105
	(0.00724)	(0.0119)	(0.00951)
ΔCurrentAccountHoldings	-0.0191**	-0.00933	-0.0174***
	(0.00638)	(0.00529)	(0.00618)
NonFinancialStockReturns	-0.123**	0.0178	-0.126***
	(0.0390)	(0.115)	(0.0273)
Constant	-0.0272	0.0848*	
	(0.0304)	(0.0454)	
Observations	18,208	18,208	18,208
R-squared	0.176	0.296	
Number of Countries	9	9	9
Time & Country FE	Yes	Yes	Yes
Chi^2 between OLS and IV of ΔNationalBankDistress			3.89**

This table shows the effects of ΔNationalBankDistress on ΔSovereignDistress for 9 Eurozone countries (Austria, Belgium, France, Greece, Ireland, Italy, Netherlands, Portugal, Spain) during the Eurozone crisis from 01/01/2009 to 12/31/2016. ΔSovereignDistress is the daily change in the natural logarithm of a country's 10-year government bond index relative to Germany's respective bond index change. ΔNationalBankDistress are asset-weighted bank stock returns on the country-level minus asset-weighted bank stock returns on the Eurozone-level. Estimated coefficients in (1) are from least squares regression. Column (2) instruments ΔNationalBankDistress with weighted stock market returns from non-Eurozone countries that are weighted according to the BIS claims of the Eurozone country towards all borrowers in the respective non-Eurozone country (NonEZStockReturns). Column (3) shows the 2nd stage of this IV regression in which ΔNationalBankDistress refers to the predicted values from (2). All variables are standardized and winsorized at the 1st and 99th percentile. All columns include country and time fixed effects on the quarterly level and the daily control variables discussed in Table 2.4 and additionally ΔUSCorporateSpread (daily change in the spread between the US corporate benchmark BBB 10-year yield and the respective AAA yield) and ΔNonFinancialItraxx (the residuals from a regression of the daily change in the natural logarithm of the 10-year Itraxx Europe against the corresponding change of the 10-year Itraxx senior and subordinated financial indices). Chi^2 shows the test statistic for the statistical difference between ΔNationalBankDistress in the OLS and the IV estimation. Standard errors (in parentheses) are clustered at the country level, ***, ** and * indicate statistical significance at the 1%, 5% and 10% level, respectively. See Table A.2.10 for variable definitions and sources.

We hypothesize that the distress transfer is stronger for countries with weaker macroeconomic performances, larger vulnerabilities in their market for sovereign bonds, impaired banking sectors and elevated political risks. For each channel under investigation we estimate the following model:[19]

$$\Delta SovereignDistress_{it} = \lambda_1 \Delta National\widehat{Bank}Distress_{it} * Channel_{it-1} + \lambda_2 Channel_{it-1} +$$
$$\lambda_3 \Delta National\widehat{Bank}Distress_{it} + \lambda_x \Delta Controls_{(i)t} + \alpha_i + \delta_t + \epsilon_{it}$$
$$(8)$$

Each interaction enters with a lag of one quarter or one month, corresponding to the frequency of the interaction variable, except when stated otherwise. This step is to account for the presumption that financial market participants in time t base their analysis on the released data from $t-1$. Concerning the sign of the interaction terms, we expect variables which increase the transmission of distress to have a positive, while factors that lower the distress transfer to have a negative coefficient.

Following Nizalova & Murtazashvili (2016) and Bun & Harrison (2019), we argue that our interaction coefficients are consistently estimated and allow for causal interpretation, as long as one variable in the interaction term is exogenously determined. We have demonstrated this assumption for our instrument and showed that potentially biasing effects are small in size and do not disturb our main results. Therefore, even if some channels could be endogenous with respect to sovereign creditworthiness, we argue that the interaction terms allow for an exogenous interpretation. We trim the interaction variables at the 1st and 99th percentile to account for potential outliers, such as Ireland's 25% increase in GDP in the first quarter of 2015 resulting from foreign companies switching their base to Ireland. Summary statistics on all non-standardized interaction variables are in Table A.2.2. All data sources are reported in the appendix in Table A.2.10.

---

[19]In practice, we estimate an IV regression with two endogenous variables (national-specific bank distress and the interaction of bank distress with the interaction term), and two instruments (international exposure-weighted stock returns and the interaction of international stock returns with the interaction term), see Wooldridge (2010) chapter 9.

Table A.2.2: Summary statistics of all interaction variables

Variable	Obs.	Mean	Median	Std. Dev.	Min	Max
DebtToGDP	21,793	98.82	96.50	29.18	48.90	179.7
FiscalDeficitToGDP	22,902	4.965	3.900	4.142	-1	28.38
GDPGrowth	22,581	0.269	0.435	1.172	-3.578	4.757
UnemploymentRate	22,508	11.09	9.200	5.800	4.300	27.40
CurrentAccountToGDP	22,640	-0.302	-0.290	5.001	-12.69	12.71
Inflation	22,945	1.391	1.400	1.453	-2.500	4.800
DebtIssuanceToGDP	16,342	2.225	2.200	1.211	0.0800	6.070
DebtRedemptionToGDP	16,337	2.007	1.990	1.244	0.0900	6.390
HighIssuance	16,641	0.498	0	0.500	0	1
HighRedemption	16,641	0.497	0	0.500	0	1
HomeShare	22,621	0.675	0.690	0.241	0.107	0.987
NonPerformingLoanRatio	20,160	8.775	5.170	7.866	2.250	36.99
ReturnOnAssets	22,448	0.184	0.330	0.670	-2.700	2.110
Tier1Ratio	21,402	12.12	11.78	3.091	7.470	25.05
BankAssetsToRevenue	22,575	33,008	28,310	19,857	16,921	120,430
CentralBankFundingShare	22,641	5.977	3.015	7.050	0.357	37.82
NonFinancialSecuritiesToGDP	22,423	8.22e-05	7.31e-05	6.33e-05	2.91e-07	0.000240
MacroprudentialIndex	18,792	0.173	-1	1.622	-2	6
PolicyUncertaintyEurope	21,717	175.5	164.4	60.58	91.38	433.3
PolicyUncertaintyEuro	22,509	0.128	0.0716	1.187	-2.402	3.939
Election	23,490	0.0239	0	0.153	0	1
Left/RightPreference	23,490	5.507	5.516	1.318	3.269	7.597
State/MarketPreference	23,490	5.587	5.241	1.327	3.105	7.913
Contra/ProEUPreference	23,490	7.953	7.903	1.071	4.758	9.253

Sample period is 01/01/2009–31/12/2016. All variables are in levels, trimmed at the 1st and 99th percentile and depicted in non-standardized format. See Table A.2.10 for definitions and sources.

A.2.1.1 Macroeconomic Performance

We test the impact of the macroeconomic performance of a Eurozone country on the isolated bank-to-sovereign distress channel using six variables. The public debt-to-GDP ratio of a country should approximate the fiscal space a government might have to finance financial sector rescue packages in times of banking sector turmoil. Similarly, the fiscal deficit to GDP ratio shows how much a government has to go into debt in a specific period and thus also approximate its potential access to financial markets in case of increasing banking sector distress. We hence expect that countries with higher debt-to-GDP ratios and larger fiscal deficits should be associated with a stronger bank-to-sovereign distress channel, i.e. a positive interaction coefficient.

Higher GDP growth and current account balance to GDP ratios as well as a lower unemployment rate should signal improved macroeconomic fundamentals and thus a reduction in the fragility transmission. We also test for the interactive effects of inflation, for which both increased or decreased transmissions are plausible.

The results in Table A.2.3 seem to confirm most of our hypotheses. We find evidence that both high public debt ratios and fiscal deficits show a positive and statistically significant relationship on the bank-to-sovereign distress channel, i.e. the transmission of distress strengthens with increasing levels of public indebtedness (columns (1) and (2)). However, both size and significance are larger for the debt-to-GDP ratio, probably because fiscal deficits were often adjusted in the short- to medium-run as a response to the crisis, while the outstanding indebtedness of the sovereign can only be reduced in the long-run.[20]

We find no statistically significant coefficient of the interaction term featuring lagged GDP growth, though it has the expected negative sign (column (3)). However, the margin plot in Figure A.2.1 provides some evidence that countries with a higher quarterly growth rate, in our case at around 1%, are no longer subject to a statistically significant transmission of distress from banks on sovereigns. Regarding the unemployment ratio, we find a highly significant positive interaction coefficient (column (4)) which suggests that periods of depressed economic performance are associated with stronger private-to-public distress transfers.

Also, countries with an increasing current account surplus seem to have statistically significant weaker transmissions of distress (column (5)), possibly indicating that economies with stronger export sectors and less import dependence seem more robust in fending off financial shocks. We cannot reject the null-hypothesis that higher inflation has no statistically significant impact on the effect of financial sector distress on sovereign creditworthiness, in accordance with our hypothesis (column (6)).

Considering the marginal effect of instrumented bank distress on sovereign bond spreads, conditional on the macroeconomic performance of a country, the marginal effect plots depicted in Figure A.2.1 support the evidence we gained in the regression framework. The marginal effect of bank distress only becomes insignificant if macroeconomic factors are sufficiently stable, in our case at a debt-to-GDP ratio of roughly 90%, a fiscal deficit under -2% of GDP, quarterly GDP growth of 1%, an unemployment ratio of 8% or a current account ratio of roughly 3%. Regarding the marginal effects of bank distress conditional on inflation, we find evidence that higher rates of inflation can indeed lead to statistically significant transmissions, however, this result seems to be primarily driven by the fewer observations of periods with lower inflation.

---

[20]At least in the absence of sovereign insolvency regimes or haircuts on government debt, which, during the crisis, was only applied for Greece.

Table A.2.3: Drivers of bank-to-sovereign distress transmissions: macroeconomic performance

	(1) ΔSovereignDistres	(2) ΔSovereignDistres	(3) ΔSovereignDistres	(4) ΔSovereignDistres	(5) ΔSovereignDistres	(6) ΔSovereignDistres
ΔNationalBankDistress	0.249*** (0.0496)	0.177*** (0.0264)	0.166*** (0.0239)	0.231*** (0.0619)	0.150*** (0.0182)	0.195*** (0.0370)
ΔNationalBankDistress × DebtToGDP	0.250*** (0.0443)					
DebtToGDP	-0.0558** (0.0244)					
ΔNationalBankDistress × FiscalDeficitToGDP		0.0919* (0.0510)				
FiscalDeficitToGDP		-0.00561 (0.0138)				
ΔNationalBankDistress × GDPGrowth			-0.0687 (0.0585)			
GDPGrowth			-0.0110* (0.00653)			
ΔNationalBankDistress × UnemploymentRate				0.286*** (0.0591)		
UnemploymentRate				-0.120*** (0.0367)		
ΔNationalBankDistress × CurrentAccountToGDP					-0.153*** (0.0338)	
CurrentAccountToGDP					-0.0701** (0.0297)	
ΔNationalBankDistress × Inflation						0.0474 (0.0428)
Inflation						0.0328 (0.0300)
Observations	18,017	17,951	17,960	17,874	17,958	17,916
Time & Country FE	Yes	Yes	Yes	Yes	Yes	Yes
Controls	Yes	Yes	Yes	Yes	Yes	Yes

This table shows the IV-regression in which ΔNationalBankDistress, which is instrumented with exposure-weighted non-Eurozone stock market returns, is interacted with quarterly or monthly variables representing the macroeconomic performance of a country, and regressed on ΔSovereignDistress. All adjustments to Table 2.4 are discussed at the beginning of this appendix chapter. The model is estimated using IV-2SLS with two endogenous variables (country-specific bank distress and the interaction between bank distress and the interaction term) and two instruments (international exposure-weighted stock returns and the interaction of international stock returns with the interaction term) for 9 Eurozone countries during the Eurozone crisis from 01/01/2009 to 12/31/2016. Each interaction term enters with a lag of one quarter or one month, depending on its frequency. DebtToGDP (1) is the debt of the general government in relation to GDP. FiscalDeficitToGDP (2) is the net borrowing (+) or net lending (-) of a general government in proportion to GDP. GDPGrowth (3) is the quarterly growth of GDP in market prices. UnemploymentRate (4) is the harmonized, seasonally adjusted unemployment rate of a country. CurrentAccountToGDP (5) is the current account of a country in ratio to its GDP. Inflation (6) is the annual rate of change in the harmonized index of consumer prices. All variables are standardized. All columns include country and time fixed effects on the quarterly level and the daily control variables discussed in Table 2.4 and additionally ΔUSCorporateSpread and ΔNonFinancialItraxx. Standard errors (in parentheses) are clustered at the country level, ***, ** and * indicate statistical significance at the 1%, 5% and 10% level, respectively. See Table A.2.10 for variable definitions and sources.

Figure A.2.1: Marginal effects of instrumented bank distress on sovereign distress conditional on macroeconomic factors. Bars indicate 95% confidence intervals. Distribution of interaction variable is shown. The results of the corresponding regressions are in Table A.2.3

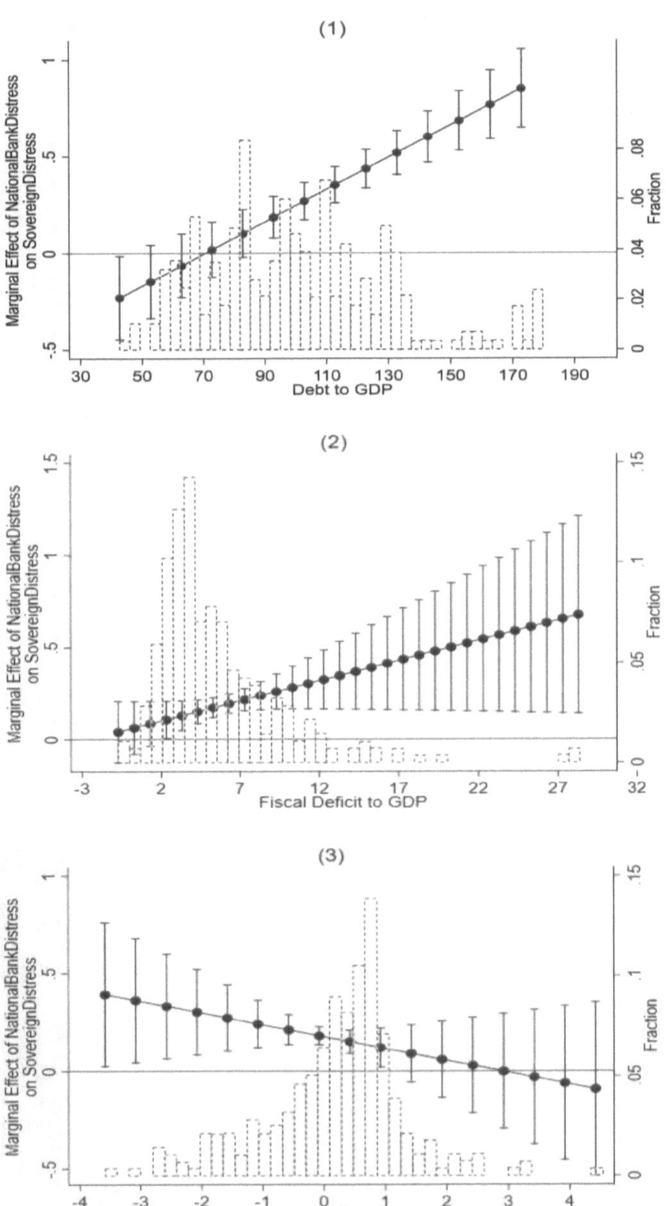

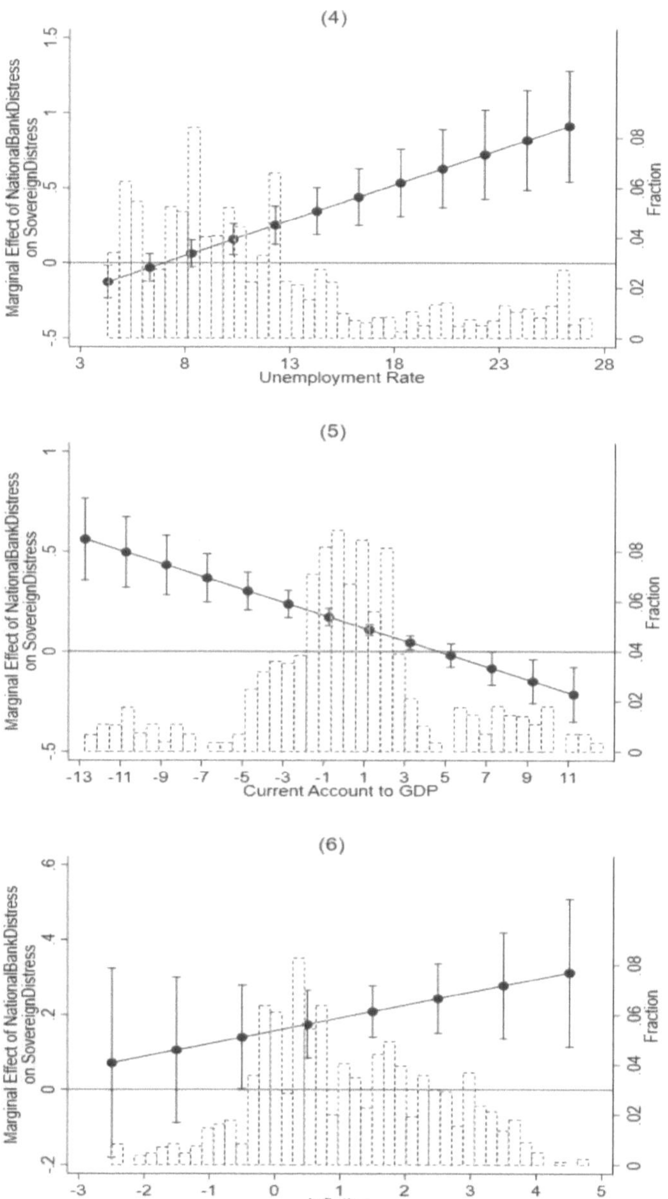

(4)
Marginal Effect of NationalBankDistress
on SovereignDistress
Fraction
Unemployment Rate
(5)
Marginal Effect of NationalBankDistress
on SovereignDistress
Fraction
Current Account to GDP
(6)
Marginal Effect of NationalBankDistress
on SovereignDistress
Fraction
Inflation

A.2.1.2 Government Bond Issuances, Redemptions and Holdings

The primary issuance of government bonds features prominently in the literature as a key transmission channel for the sovereign-bank loop. Both Gaballo & Zetlin-Jones (2016) and Farhi & Tirole (2017) model the loop with a sovereign that issues new public debt in order to finance bank bailouts. The increased supply of bonds lowers their prices and raises their interest rates. Bond losses are transmitted to bank balance sheets that prefer holding domestic sovereign debt which necessitates further bailouts. Similarly, Ongena et al. (2019) identify months in which governments have to roll-over maturing government debt as periods in which the sovereign likely performs moral suasion towards its banking sector, pressuring domestic banks to stand ready as buyers of government debt. Following this literature, it is likely that periods with higher issuances or redemptions of government bonds could be associated with a stronger transmission of bank-to-sovereign distress.

We therefore test if the effect of instrumented bank distress on sovereign bond spreads is conditional on the amount of government debt a country issues or repays. We assume that the effect of the issuance or redemption should affect the financial distress of the same month, hence we consecutively interact the contemporaneous level of issuance or redemption of government bonds in relation to a country's GDP with instrumented bank distress in our estimation. In both cases, we find a similar effect on the transmission of distress that is, however, statistically indistinguishable from zero, as shown in columns (1) and (2) of Table A.2.4. Concerning the marginal effect of the distress channel conditional on government bond issuances and redemptions in Figure A.2.2, we find that there are values of the interaction variable for which the marginal effect of bank distress turns insignificant, however, this finding seems to be clearly driven by fewer observations and therefore wider confidence intervals in the distribution of government bond redemptions or issuances.

We further investigate this channel by using the actual amounts of government bonds issued or redeemed and marking the months in which the amounts of bonds given out or repaid lay above the median for this country during our sample with a dummy that is 1 in these months and 0 otherwise. Again, we find no statistically significant effects when we interact this variable with the instrumented bank distress measure (columns (3) and (4)). We view this finding as evidence that, while there may certainly have been cases in which large chunks of new government bonds entering the market could have had price effects, financial markets, on average, did not discriminate between months with high or low government debt issuances when they re-priced such securities.

Lastly, we look if the holding of domestic government bonds by the banking sector could have had an impact on the bank-sovereign distress channel. The literature has shown that governments likely pressure domestic banking sectors to purchase their own securities to ease refinancing (De Marco & Macchiavelli (2016), Ongena et al. (2019)). A banking sector with strong exposure towards its own government getting hit by a negative shock could have a detrimental impact on sovereign creditworthiness as the bank might rapidly sell government securities to raise liquidity. Another possible channel is that the holding of government bonds artificially increases the equity ratio of the bank due to the zero risk weight of government securities which makes the bank look safer on paper than in practice. Lastly, the bank might be more likely to be bailed out because the government wants to keep it as a buyer of government securities. Consequently, we interact our main specification with the lagged home share of government bonds by a banking sector, i.e. the share of domestic government debt securities held compared to the total holdings of government securities. Column (5) in Table A.2.4 and the margin plot in Figure A.2.2 show that a larger home share has a strong and statistically highly significant positive effect on the bank-sovereign distress channel, confirming the direction of the discussed channels.

Table A.2.4: Drivers of bank-to-sovereign distress transmissions: issuance, redemption and holding of government debt

	(1) ΔSovereignDistres	(2) ΔSovereignDistres	(3) ΔSovereignDistres	(4) ΔSovereignDistres	(5) ΔSovereignDistres
ΔNationalBankDistress	0.269***	0.266***	0.276***	0.264***	0.230***
	(0.0513)	(0.0499)	(0.0360)	(0.0470)	(0.0509)
ΔNationalBankDistress	0.00115				
$\times$ DebtIssuanceToGDP	(0.0652)				
DebtIssuanceToGDP	-0.00764				
	(0.0112)				
ΔNationalBankDistress		0.00636			
$\times$ DebtRedemptionToGDP		(0.0510)			
DebtRedemptionToGDP		0.00497**			
		(0.00234)			
ΔNationalBankDistress			-0.0117		
$\times$ HighIssuance			(0.0707)		
HighIssuance			-0.0316		
			(0.0212)		
ΔNationalBankDistress				0.0117	
$\times$ HighRedemption				(0.0346)	
HighRedemption				0.0379**	
				(0.0177)	
ΔNationalBankDistress					0.206***
$\times$ HomeShare					(0.0694)
HomeShare					0.0680
					(0.0743)
Observations	15,824	15,823	16,116	16,116	17,872
Time & Country FE	Yes	Yes	Yes	Yes	Yes
Controls	Yes	Yes	Yes	Yes	Yes

This table shows the IV-regression in which ΔNationalBankDistress, which is instrumented with exposure-weighted non-Eurozone stock market returns, is interacted with quarterly or monthly variables representing the issuance, redemption or holding of government debt securities, and regressed on ΔSovereignDistress. All adjustments to Table 2.4 are discussed at the beginning of this appendix chapter. The model is estimated using IV-2SLS with two endogenous variables (country-specific bank distress and the interaction of bank distress with the interaction term) and two instruments (international exposure-weighted stock returns and the interaction of international stock returns with the interaction term) for 9 Eurozone countries during the Eurozone crisis from 01/01/2009 to 12/31/2016. The variables enter with their contemporaneous value unless stated otherwise. DebtIssuanceToGDP (1) is the gross issuance of debt securities of a general government in relation to GDP. DebtRedemptionToGDP (2) is the gross redemption of debt securities of a general government in relation to GDP. HighIssuance (3) and HighRedemption (4) are dummy variables equal to one in months in which the gross issuance (3) or gross redemption (4) of government debt was above the sample median of this country. HomeShare (5) is the share of domestic government bonds held by a country's banking sector compared to its total government bond holding. All non-dummy variables are standardized. All columns include country and time fixed effects on the quarterly level and the daily control variables discussed in Table 2.4 and additionally ΔUSCorporateSpread and ΔNonFinancialItraxx. Standard errors (in parentheses) are clustered at the country level, ***, ** and * indicate statistical significance at the 1%, 5% and 10% level, respectively. See Table A.2.10 for variable definitions and sources.

Figure A.2.2: Marginal effects of instrumented bank distress on sovereign distress conditional on issuance, redemption or holdings of government debt securities. Bars indicate 95% confidence intervals. Distribution of interaction variable is shown. The results of the corresponding regressions are in Table A.2.4

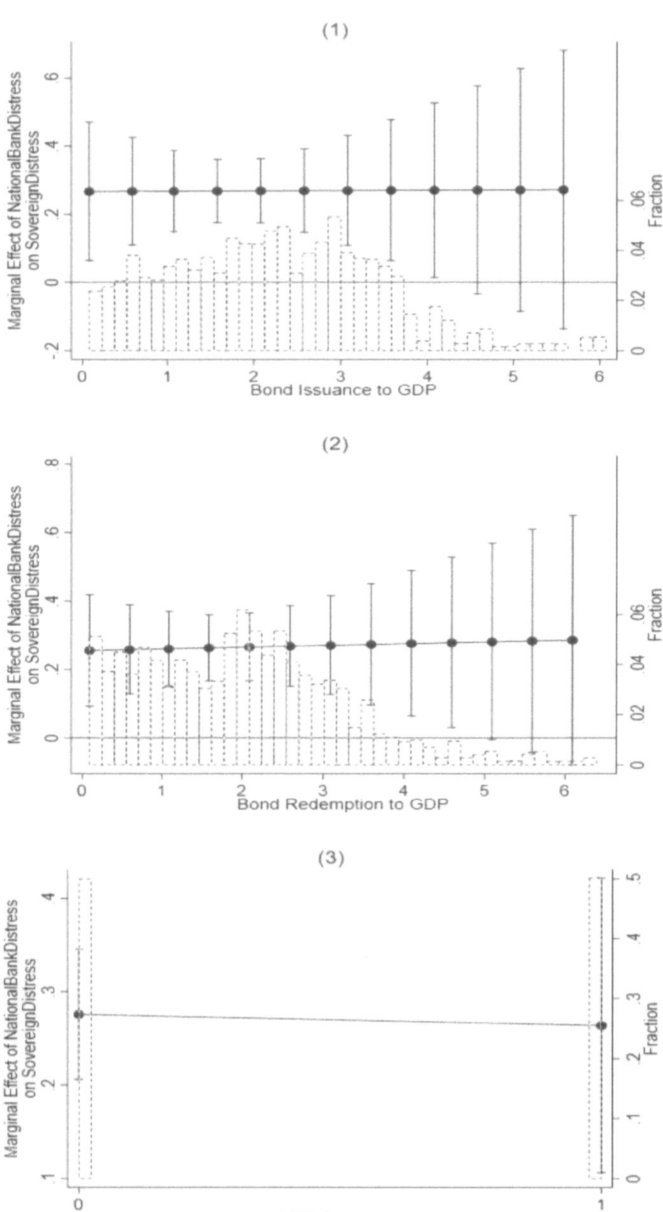

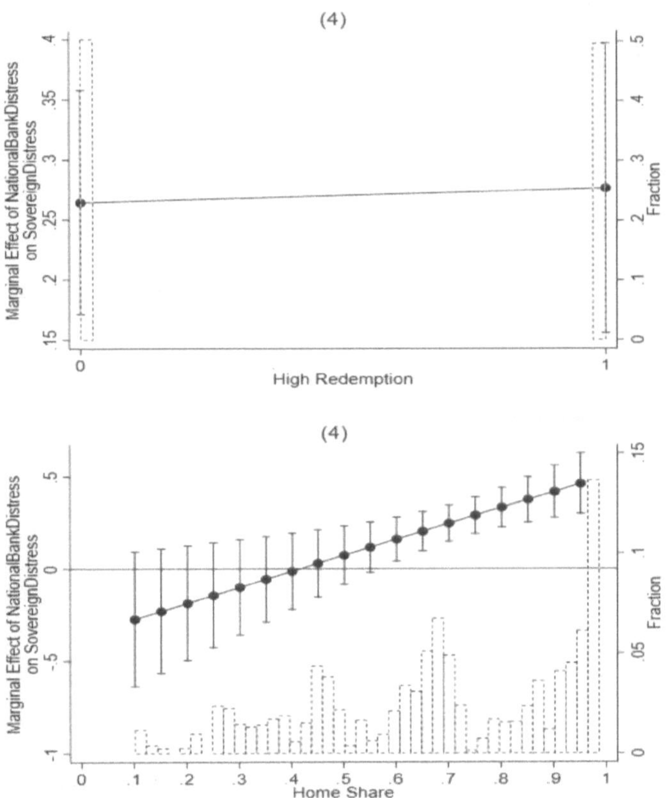

A.2.1.3 Banking Sector Structure and Stability

In the following, we shed light on the link between the bank-to-sovereign distress channel and the structure and stability of the corresponding country's banking sector. Our hypothesis is that the effect of bank distress, instrumented by imported stock market returns, is conditional on the profitability, capitalization, amount of non-performing loans, liability structure and size of the banking sector. Furthermore, we suspect that both stronger macroprudential regulation and more developed capital markets can cushion the transmission of distress, as the former might point to a more comprehensive regulatory handling of financial sector shocks whereas the latter can serve as a substitute for firm financing in case of an impaired banking sector.

We interact the instrumented bank distress measure consecutively with these variables and report results in Table A.2.5. The findings suggest that the transmission of distress depends in a statistically significant manner on the return on assets (column (2)) and the non-performing loans ratio (column (1)) of a banking sector, the former with a cushioning impact when higher, the latter with an accelerating one, as hypothesized. With regard to the interaction of the Tier 1 capital ratio, we find a negative effect that does, however, not differ statistically significant from zero (column (3)). This finding could be explained by the fact that higher capital requirements were arguably the most often prescribed action by regulators for ailing banking sectors. Demanding higher equity in times when the transmission of bank distress is strongest would bias the corresponding coefficient towards zero and hence account for the result. However, when investigating the marginal effects depicted in Figure A.2.3, we find indeed some evidence suggesting that banking sectors with sufficient Tier 1 capital ratios, in our case at around 16%, are no longer subject to a statistically significant marginal effect of bank distress on sovereign creditworthiness. The marginal effects of bank distress, conditional on non-performing loans or return on assets are, in a similar vein, pointing towards an insignificant transmission of distress when non-performing loans are lower than 4% and return on assets higher than 0.8%. Stronger banking sectors, in terms of capitalization, non-performing loans or profitability, may therefore contribute to less distress spillovers on sovereign creditworthiness.

Turning towards the size of the banking sector, we interact national bank distress with the total bank asset to public revenue ratio of a country. This ratio sets the size of banks in relation to the fiscal means the government has in this period to potentially finance rescue packages. However, we cannot statistically significantly reject the null hypothesis of a zero effect of this interaction term on sovereign creditworthiness (column (4) in Table A.2.5). One interpretation for this result could be that too-big-to-fail banking sectors are not an exclusive driver of the bank-sovereign distress channel but that interconnected financial sectors, for instance in the case of a regional banking system in distress as witnessed in Italy or Spain, can also be responsible for spreading financial distress to the sovereign, even if they are smaller in size.

Next, we interact bank distress with a measure for liability risks of the financial sector. We use the share of banking sector liabilities that is funded by the central bank, i.e. the ECB or in practice the national central bank. Banks that turn to the central bank to finance their assets likely do so, because it is more expensive or no longer possible for them to receive funds on private markets. Indeed, during the Eurozone crisis international money market funds started to withdraw short-term funding for Eurozone banks in 2011, with the ECB stepping in as a lender of last resort to limit the funding gap (Acharya, Pierret & Steffen (2018)). In our estimation, the positive and highly statistically significant interaction term in column (5) suggests that banking sectors that required more financing by the central bank also featured a stronger transmission of bank-to-sovereign distress.

Finally, we investigate the impact of two measures that could potentially cushion the analyzed distress channel. We obtain the cumulative macroprudential index from Cerutti et al. (2016) which is an index that sums up all macroprudential instruments such as sector-specific capital buffers or loan-to-value caps introduced by regulators on a quarterly frequency. Also, we estimate the model using the amount of debt securities issued by non-financial firms in relation to GDP as an interaction term. This variable approximates how well firms could substitute bank credit in case the loan supply by banks was disrupted during the crisis. We find negative but not statistically significant effects for both interaction terms (columns (6) and (7)). However, the marginal effect plots in Figure A.2.3 provide some evidence that countries with higher macroprudential regulation or a more pronounced capital market are subject to a lower and at some point statistically insignificant transmission of bank to sovereign distress. These results suggest that macroprudential regulation and developed capital markets are not primary channels for the bank-to-sovereign connection, but may cushion the transmission if they are developed to a critical level.

	(1) ΔSovereignDistres	(2) ΔSovereignDistres	(3) ΔSovereignDistres	(4) ΔSovereignDistres	(5) ΔSovereignDistres	(6) ΔSovereignDistres	(7) ΔSovereignDistres
ΔNationalBankDistress	0.256*** (0.0462)	0.180*** (0.0310)	0.191*** (0.0499)	0.205*** (0.0461)	0.188*** (0.0523)	0.176*** (0.0379)	0.214*** (0.0419)
ΔNationalBankDistress × NonPerformingLoanRatio	0.312*** (0.0403)						
NonPerformingLoanRatio	-0.114*** (0.0315)						
ΔNationalBankDistress × ReturnOnAssets		-0.108** (0.0457)					
ReturnOnAssets		0.0140 (0.0113)					
ΔNationalBankDistress × Tier1Ratio			-0.0182 (0.0344)				
Tier1Ratio			-0.00310 (0.0209)				
ΔNationalBankDistress × BankAssetsToRevenue				-0.0202 (0.0476)			
BankAssetsToRevenue				0.0245** (0.00969)			
ΔNationalBankDistress × CentralBankFundingShare					0.288*** (0.0425)		
CentralBankFundingShare					-0.0536*** (0.00877)		
ΔNationalBankDistress × NonFinancialSecuritiesToGDP						-0.140 (0.112)	
NonFinancialSecuritiesToGDP						0.0579 (0.0409)	
ΔNationalBankDistress × MacroprudentialIndex							-0.0579 (0.0421)
MacroprudentialIndex							0.0132 (0.00834)
Observations	17,956	18,021	17,955	17,869	17,936	17,616	14,265
Time & Country FE	Yes	Yes	Yes	Yes	Yes	Yes	Yes
Controls	Yes	Yes	Yes	Yes	Yes	Yes	Yes

This table shows the IV-regression in which ΔNationalBankDistress, which is instrumented with exposure-weighted non-Eurozone stock market returns, is interacted with quarterly or monthly variables representing the banking sector structure and stability of a country, and regressed on ΔSovereignDistress. All adjustments to Table 2.4 are discussed at the beginning of this appendix chapter. The model is estimated using IV-2SLS with two endogenous variables (country-specific bank distress and the interaction of bank distress with the interaction term) and two instruments (international exposure-weighted stock returns and the interaction of international stock returns with the interaction term) for 9 Eurozone countries during the Eurozone crisis from 01/01/2009 to 12/31/2016. Each interaction term enters with a lag of one quarter or one month, depending on its frequency. NonPerformingLoanRatio (1) is the ratio of non-performing loans to total gross loans of a banking sector. ReturnOnAssets (2) is the return on assets ratio of a banking sector. Tier1Ratio (3) is the ratio of regulatory Tier1 capital to risk-weighted assets. BankAssetsRevenue (4) is the ratio of the total balance sheet size of a banking sector towards the total general government's revenues. CentralBankFundingShare (5) is the share of liabilities in a country's banking sector funded by the central bank. NonFinancialSecuritiesToGDP (6) is the ratio of outstanding securities issued by non-financial corporations in ratio to GDP. MacroprudentialIndex (7) is an index from Cerutti et al. (2016) that aggregates the number of implemented macroprudential instruments in a country. All non-index variables are standardized. All columns include country and time fixed effects on the quarterly level and the daily control variables discussed in Table 2.4 and additionally ΔUSCorporateSpread and ΔNonFinancialItraxx. Standard errors (in parentheses) are clustered at the country level, ***, ** and * indicate statistical significance at the 1%, 5% and 10% level, respectively. See Table A.2.10 for variable definitions and sources.

Figure A.2.3: Marginal effects of instrumented bank distress on sovereign distress conditional on banking sector stability. Bars indicate 95% confidence intervals. Distribution of interaction variable is shown. The results of the corresponding regressions are in Table A.2.5

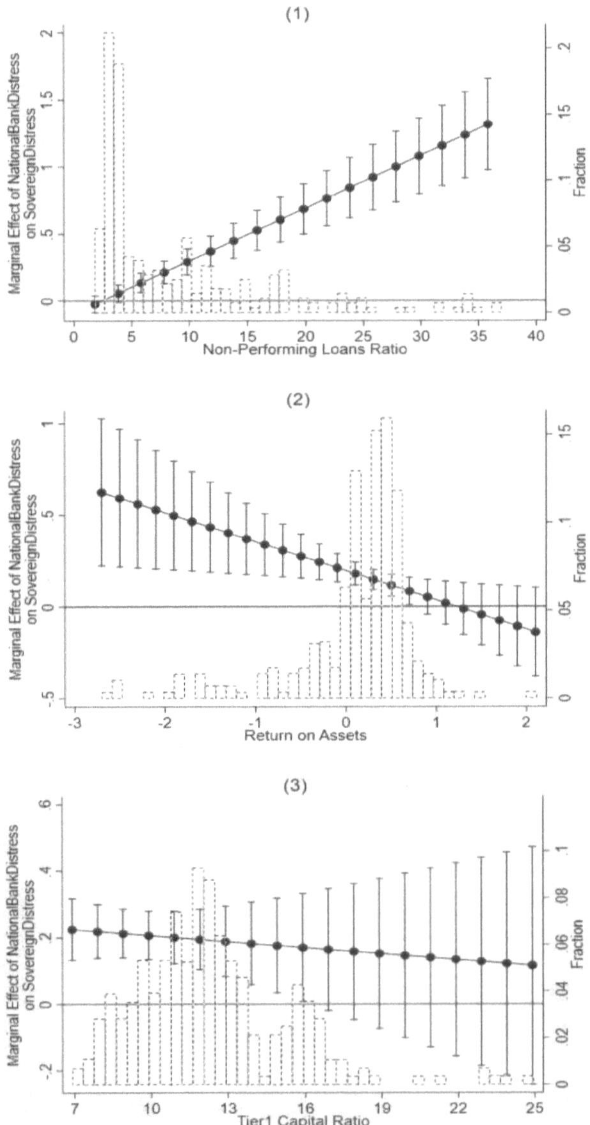

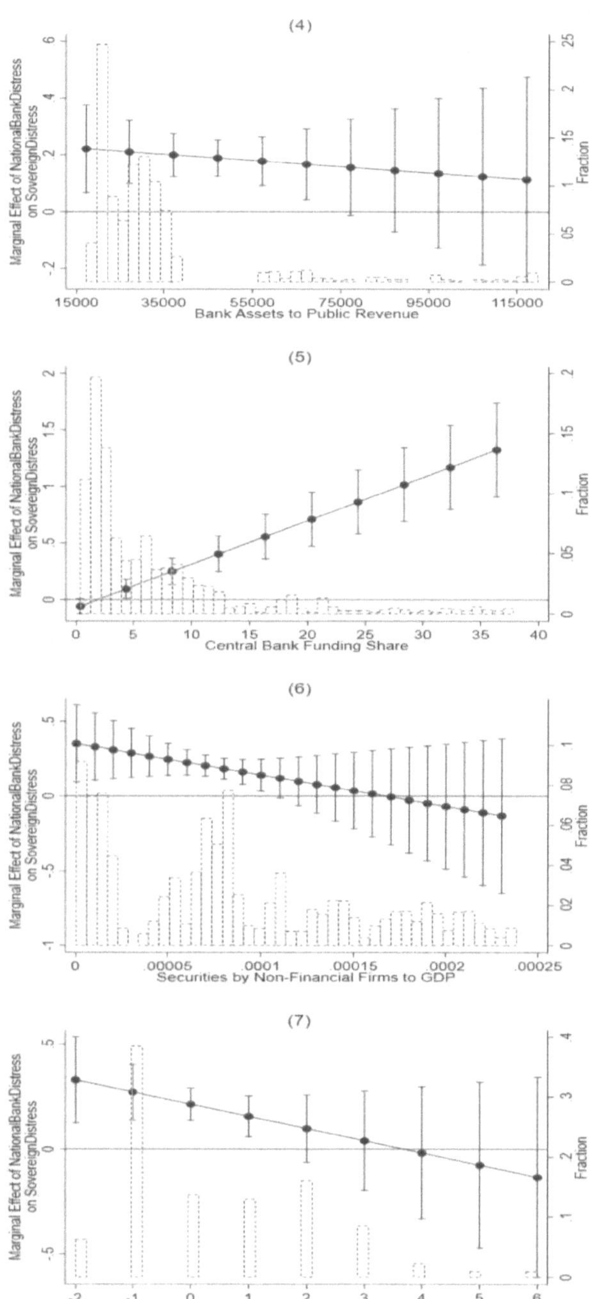

(4)
Marginal Effect of NationalBankDistress on SovereignDistress
Fraction
Bank Assets to Public Revenue
(5)
Marginal Effect of NationalBankDistress on SovereignDistress
Fraction
Central Bank Funding Share
(6)
Marginal Effect of NationalBankDistress on SovereignDistress
Fraction
Securities by Non-Financial Firms to GDP
(7)
Marginal Effect of NationalBankDistress on SovereignDistress
Fraction
Macroprudential Regulation Index

A.2.1.4 Political Stability

Lastly, we test whether different levels of political risk or events approximating them have a significant effect on bank-sovereign distress transfers. We hypothesize that elevated political uncertainty in Europe or the Eurozone could increase the transmission of distress because they make a collaborative approach concerning the regulatory architecture of the Eurozone or a common political strategy on rescue packages for banks or countries more difficult. This insecurity could feed into the bank-sovereign distress channel by creating a lack of clarity or ambiguity in handling financial sector shocks which ultimately leaves sovereigns and taxpayers on the country-level to deal with these risks.

To analyze this channel, we interact instrumented bank distress with a political uncertainty index for Europe as established by Baker et al. (2016). This continuous index is based on articles from various European newspapers covering political uncertainty and hence provides a monthly-varying approximating of political risks in Europe. The results in Table A.2.6 suggest that an increasing level of political uncertainty can lead to a stronger transmission of bank distress on sovereign creditworthiness, as the interaction term is positive and statistically highly significant (column (1)).

In order to test if political risks that are more closely related to the Euro Area crisis than the broader European index have a similar effect, we proceed as follows: We collect the political uncertainty indices for Ireland, Spain and Italy which are the only GIIPS countries with an uncertainty index. We then conduct a principal component analysis and estimate the first component of these three indices. This component describes variation in the indices that is common for all three countries and should therefore pick up political risk factors that are shared by these countries, such as uncertainty related to the future of the EZ. This first component accounts for 57.77% of the total variation. Interacting the variable with national-specific bank distress, we find a positive statistically significant effect at the 1% level for the interaction term (column (2)).

Next, we test the effect of political uncertainty on the country level. We use the parlgov database to pinpoint the months in which federal parliamentary elections for each country in our panel took place. We give these months a value of 1 and 0 for months without elections. Interacting this dummy with our measure for bank distress, we find a highly positive and statistically significant effect which suggests that months with elections and hence greater political uncertainty seem to be associated with a stronger bank-sovereign distress channel compared to months without elections (column (3)). These results suggest that in times of elevated political uncertainty, the transmission of distress from banking sectors to sovereigns is amplified.

Lastly, we exploit the parlgov database to test if the political preferences of political parties in ruling governments had any impact on the investigated distress channel. Similar to Eichler & Sobański (2016), we weight the size of the political parties in the ruling cabinet based on their seats in parliament. We then create a weighted index of the government's stance of being a left versus a conservative, a state-friendly versus a market-friendly and a pro-European versus an EU-skeptical cabinet coalition. A higher value of the index indicates a more conservative, market-friendly or EU-friendly government respectively. However, the interaction of the contemporaneous index value with the bank distress measure yields small and in each case statistically insignificant coefficients (columns (4)-(6) of Table A.2.6). Figure A.2.4 supports this conclusion, as the marginal effects of bank distress depicted are largely unconditional towards the government's ideological stance. Only for more pro-EU governments, there seems to be a stronger transmission of distress. However, it should be noted that almost all governments during our sample period are represented by a pro- or at least EU-tolerating index which limits the informative value of this outcome. Overall, these results could suggest that the political ideology of governments during the Eurozone crisis, on average, had only a secondary role when it comes to coping with the bank-sovereign distress channel.

Table A.2.6: Drivers of bank-to-sovereign distress transmissions: political stability

	(1) ΔSovereignDistres	(2) ΔSovereignDistres	(3) ΔSovereignDistres	(4) ΔSovereignDistres	(5) ΔSovereignDistres	(6) ΔSovereignDistres
ΔNationalBankDistress	0.231*** (0.0386)	0.213*** (0.0396)	0.191*** (0.0364)	0.204*** (0.0423)	0.200*** (0.0419)	0.200*** (0.0383)
ΔNationalBankDistress × PolicyUncertaintyEurope	0.125*** (0.0344)					
PolicyUncertaintyEurope	0.00562 (0.00576)					
ΔNationalBankDistress × PolicyUncertaintyEuro		0.111*** (0.0267)				
PolicyUncertaintyEuro		-0.00653 (0.00878)				
ΔNationalBankDistress × Election			0.355*** (0.107)			
Election			0.102** (0.0508)			
ΔNationalBankDistress × Left/RightPreference				-0.0293 (0.0437)		
Left/RightPreference				-0.0272* (0.0146)		
ΔNationalBankDistress × State/MarketPreference					0.00632 (0.0397)	
State/MarketPreference					-0.0263* (0.0139)	
ΔNationalBankDistress × Conta/ProEUPreference						0.0786 (0.0554)
Conta/ProEUPreference						0.0177 (0.0114)
Observations	18,208	18,208	18,208	18,208	18,208	18,208
Time & Country FE	Yes	Yes	Yes	Yes	Yes	Yes
Controls	Yes	Yes	Yes	Yes	Yes	Yes

This table shows the IV-regression in which ΔNationalBankDistress, which is instrumented with exposure-weighted non-Eurozone stock market returns, is interacted with quarterly or monthly variables representing the political stability of a country, and regressed on ΔSovereignDistress. All adjustments to Table 2.4 are discussed at the beginning of this appendix chapter. The model is estimated using IV-2SLS with two endogenous variables (country-specific bank distress and the interaction of bank distress with the interaction term) and two instruments (international exposure-weighted stock returns and the interaction of international stock returns with the interaction term) for 9 Eurozone countries during the Eurozone crisis from 01/01/2009 to 12/31/2016. PolicyUncertaintyEurope (1) is a policy uncertainty index for Europe established in Baker et al. (2016). PolicyUncertaintyEuro (2) is the first component of a principal component analysis of the policy uncertainty indices of Ireland, Spain and Italy. Election (3) is a dummy that is 1 if the country held a parliamentary election in that month. Left/Right, State/Market and Contra/ProEU (4, 5, 6) are weighted indices of the political preferences of ruling parties in the cabinet weighted by their seats in parliament. A higher index implies a more conservative/market-friendly or EU-friendly government, respectively. The interaction terms in (1) and (2) enter with a lag of one month, whereas the other variables enter the estimation with their contemporaneous value. All non-dummy variables are standardized. All columns include country and time fixed effects on the quarterly level and the daily control variables discussed in Table 2.4 and additionally ΔUSCorporateSpread and ΔNonFinancialItraxx. Standard errors (in parentheses) are clustered at the country level. ***, ** and * indicate statistical significance at the 1%, 5% and 10% level, respectively. See Table A.2.10 for variable definitions and sources.

Figure A.2.4: Marginal effects of instrumented bank distress on sovereign distress conditional on political stability. Bars indicate 95% confidence intervals. Distribution of interaction variable is shown. The results of the corresponding regressions are in Table A.2.6

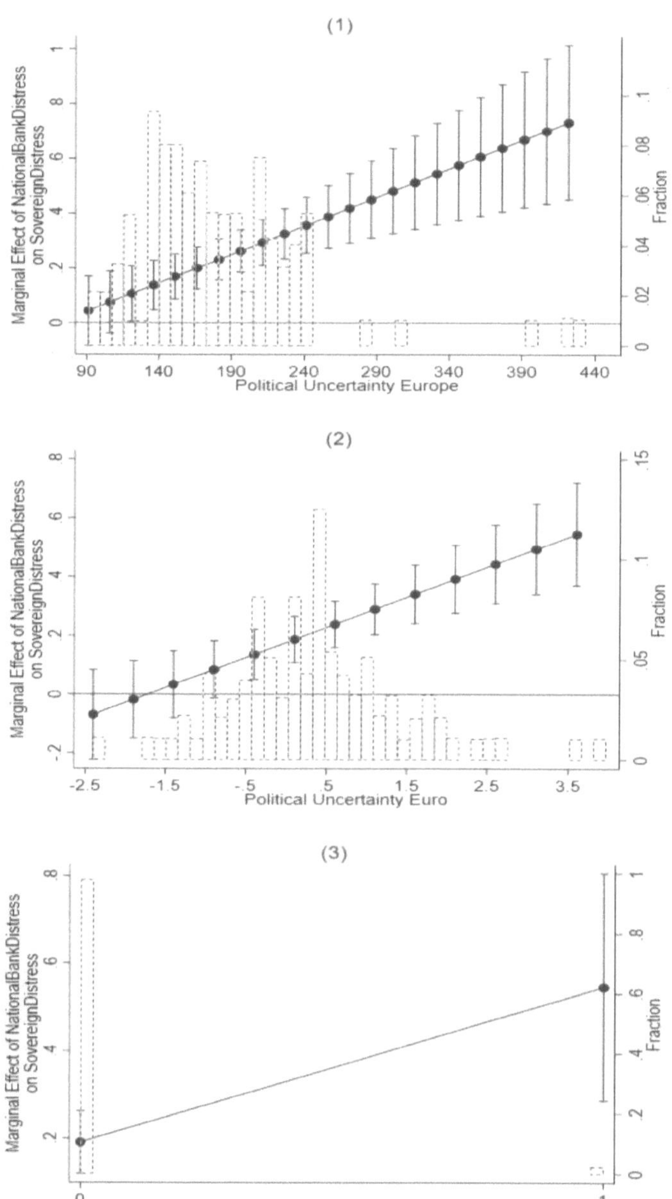

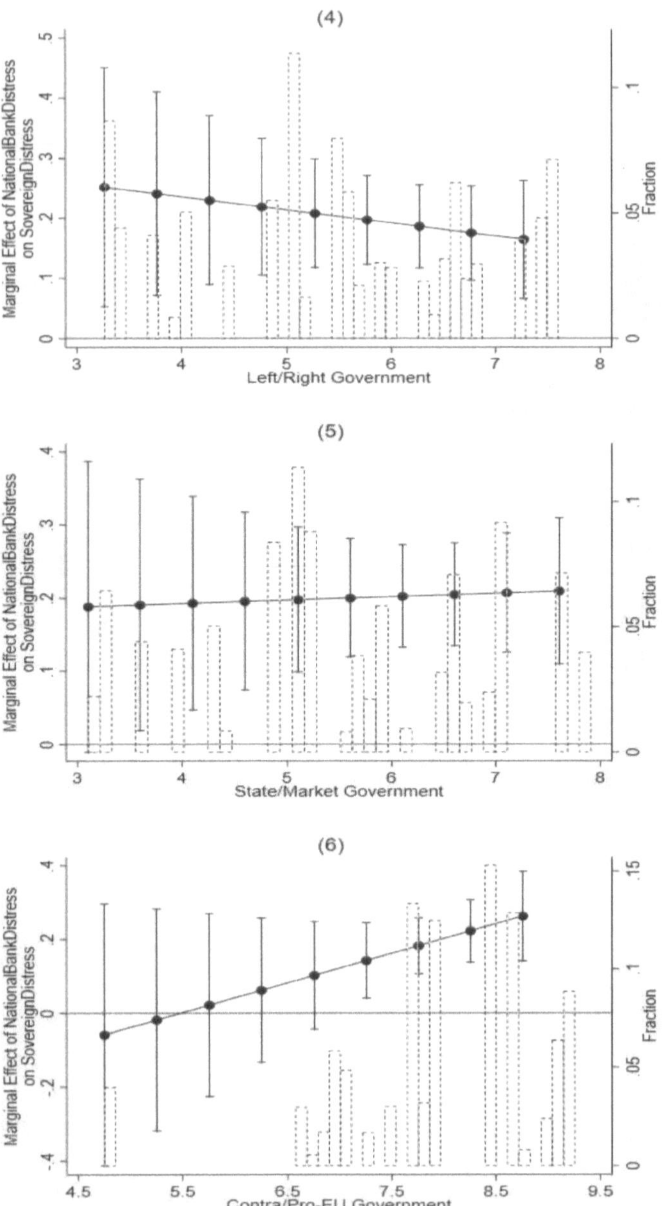
(4)
Marginal Effect of NationalBankDistress on SovereignDistress
Fraction
Left/Right Government
(5)
Marginal Effect of NationalBankDistress on SovereignDistress
Fraction
State/Market Government
(6)
Marginal Effect of NationalBankDistress on SovereignDistress
Fraction
Contra/Pro-EU Government

A.2.2 Additional Tables and Figures

Table A.2.7: List of included banks (mean asset size in Euro in brackets)

Country	Banks
Austria	BKS Bank (4.29 bn), Bank Tirol and Vorarlberg (4.47 bn), Erste Group Bank (165 bn), Oberbank (14.0 bn), Raiffeisen Bank International (104 bn)
Belgium	Dexia (433 bn), KBC Ancora (2.9 bn), KBC Group (267 bn)
France	Banque De La Reunion (2.1 bn), Banque Tarneaud (2.0 bn), BNP Paribas (1490 bn), Boursorama (2.5 bn), Credit Agricoles Alpes Provences (12.2 bn), Credit Agricole Atlantique Vendee (14.8 bn), Credit Agricole Brie Picardie (19.7 bn), Credit Agricole Centre Loire (8.9 bn), Credit Credit Agricole d'Ile de France (27.9 bn), Credit Agricole d'Ille-et-Vilaine (7.9 bn), Credit Agricole Languedoc (20.5 bn), Credit Agricole Loire Haute-Loire (7.4 bn), Credit Agricole Normandie Seine (9.9 bn), Credit Agricole Morbihan (6.9 bn), Credit Agricole Nord de France (23.8 bn), Credit Agricole SA (1360 bn), Credit Agricole Sud Rhone Alpes (11.2 bn), Credit Agricole Toulouse (6.9 bn), Credit Agricole Touraine Poitou (8.3 bn), Credit Foncier de Monaco (2.8 bn), Credit Industriel et Commercial CIC (208 bn), Natixis (360 bn), Rothschild & Co (5.4 bn), Societe Generale (935 bn)
Greece	Agricultural Bank of Greece (22.7 bn), Alpha Bank (50.1 bn), Attica Bank (3.15 bn), Bank of Piraues (43.5 bn), Emporiki Bank (21.2 bn), Eurobank Ergasias (54.7 bn), General Bank of Greece (3.4 bn), Marfin Egnatia Bank (8.4 bn), Marfin Investment Group (3.0 bn), National Bank of Greece (81.5 bn), Proton Bank (2.6 bn), T Bank (2.0 bn), TT Hellenic Postbank (14.6 bn)
Ireland	Allied Irish Banks (123 bn), Anglo Irish Bank (50.5 bn), Bank Of Ireland (140 bn), Permanent Tsb Group (50.9 bn)
Italy	Banca Carige (27.4 bn), Banca Finnat Euramerica (0.66 bn), Banca Generali (5.5 bn), Banca IFIS (3.4 bn), Banca Intermobiliare (2.8 bn), Banca Italease (21.5 bn), Banca Mediolanum (24.0 bn), Banca Monte Dei Paschi (166 bn), Banca Piccolo Credito Valtellinese (18.7 bn), Banca Popolare Italiana (38.8 bn), Banca Popolare dell'Etruria e del Lazio (8.4 bn), Banca Popolare di Milano (41.4 bn), Banca Popolare di Sondrio (21.0 bn), Banca Popolare di Spoleto (2.3 bn), Banca Profilo (1.6 bn), Banca Sistema (2.1 bn), Banco Di Sardegna (12.3 bn), Banco BPM (89.4 bn), Banco di Desio e della Brianza (7.2 bn), BPER Banca (47.7 bn), Credito Artigiano (6.5 bn), Credito Bergamasco (12.3 bn), Credito Emiliano (25.9 bn), FinecoBank (19.3 bn), Intesa Sanpaolo (484 bn), IW Bank (1.9 bn), Mediobanca (55.1 bn), Meliorbanca (3.5 bn), Unicredit (664 bn), Unione di Banche Italiane (108 bn)
Netherlands	ABN AMRO (394 bn), Binckbank (2.2 bn), ING Groep (988 bn), KAS Bank (6.2 bn), SNS Reaal (120 bn), Van Lanschot (16.3 bn)
Portugal	Banco BPI (34.9 bn), Banco Comercial Portugues (75.9 bn), Banco Espirito Santo (57.6 bn), Banif Financial Group (9.4 bn), Finibanco (2.0 bn), Montepio (21.2 bn)
Spain	Banca Civica (69.8 bn), Banco Bilbao Vizcaya Argentaria (466 bn), Banco De Andalucia (7.6 bn), Banco de Castilla (3.3 bn), Banco de Credito Balear (1.4 bn), Banco de Galicia (2.8 bn) Banco De Sabadell (99.8 bn), Banco De Valencia (14.1 bn), Banco de Vasconia (2.5 bn) Banco Espanol De Credito (84.1 bn), Banco Guipuzcoano (7.3 bn), Banco Pastor (19.1 bn), Banco Popular Espanol (98.8 bn), Banco Santander (862 bn), Bankia (228 bn), Bankinter (43.6 bn), Caixabank (231 bn), Caja De Ahorros Del Mediterraneo (54.4 bn), Liberbank (39.9 bn)

Germany: banks used in construction of *EurozoneBank–* *Returns*$_t$	Aareal Bank (42.3 bn), BHW Holding (94.4 bn), Baader Bank (0.37 bn), Bayerische Hypo- und Vereinsbank (539 bn), Berlin-Hannoversche Hypothekenbank (40.2 bn), Comdirect Bank (9.2 bn), Commerzbank (544 bn), Deutsche Bank (1410 bn), Deutsche Hypothekenbank (29.8 bn), Deutsche Pfandbriefbank (63.4 bn), Deutsche Postbank (185 bn), Eurohypo (182 bn), HSBC Trinkaus and Burkhardt (13.5 bn), Hypo Real Estate (252 bn), IKB (34.6 bn), Landesbank Berlin Holding (154 bn), Merkur Bank (0.7 bn), Net-M Privatbank (0.065 bn), Oldenburgische Landesbank (11.9 bn), Quirin Bank (4.1 bn), Umweltbank (1.7 bn), Varengold Bank (0.49 bn)

Table A.2.8: List of non-Eurozone countries used for instrument in Section 2.3.2

Argentina, Australia, Bahrain, Brazil, Bulgaria, Canada, Chile, China, Colombia, Croatia, Czech Republic, Denmark, Egypt, Hong Kong, Hungary, India, Indonesia, Israel, Japan, Jordan, Kuwait, Malaysia, Mexico, Morocco, New Zealand, Nigeria, Norway, Oman, Pakistan, Peru, Philippines, Poland, Qatar, Romania, Russia, Singapore, South-Africa, South-Korea, Sri Lanka, Sweden, Switzerland, Thailand, Turkey, United Arab Emirates, United Kingdom, United States, Venezuela

Table A.2.9: Non-Eurozone countries that are "large borrowers" of a Eurozone country as defined in Section 2.3.2

Eurozone country	"Large borrower" countries (non-Eurozone)
Austria	Bulgaria, Croatia, Czech Republic, Hungary, Romania, Russia
Belgium	Bahrain, Czech Republic, Hungary
France	Australia, Bahrain, Bulgaria, China, Colombia, Croatia, Czech Republic, Denmark, Egypt, Hong Kong, Israel, Japan, Jordan, Kuwait, Morocco, Nigeria, Oman, Pakistan, Peru, Poland, Qatar, Romania, Russia, South Korea, Sri Lanka, Sweden, Switzerland, Turkey, United Arab Emirates, United Kingdom
Greece	Bulgaria, Romania, Turkey
Ireland	
Italy	Bulgaria, Croatia, Egypt, Hungary, Peru, Poland, Russia
Netherlands	Australia, Brazil, Canada, Denmark, Poland, Romania, Russia, Sweden, United Arab Emirates, United Kingdom
Portugal	
Spain	Argentina, Brazil, Chile, Colombia, Mexico, Morocco, Peru, Poland, United Kingdom, Venezuela

Table A.2.10: Description and sources of variables

Variable	Description	Source
Bank-specific Variables (all return variables winsorized at 1st and 99th percentile)		
Weighted Bank Stock Returns	Daily change in the natural logarithm of bank stocks weighted with yearly total asset size of bank in that country. Return is set to missing within a quarter if the stock had no turnover or no stock value was reported for more than seven consecutive trading days.	Datastream
BankReturns	Asset-weighted bank stock returns aggregated on the country-level.	

World Stock Returns Banks Excluding EZ	Daily change in natural logarithm of a world stock index of bank stocks that excludes stocks from the EZ.	Datastream
Eurozone Bank Returns	Asset-weighted bank stock returns on the Eurozone-level, i.e. of all banks in the panel. Variable is orthogonalized with respect to World Stock Returns Banks (Excl. EZ).	
NationalBankDistress	BankReturns minus EurozoneBankReturns. Variable is then multiplied times -1.	

Non-Eurozone Exposure Variables (all return variables winsorized at 1st and 99th percentile)		
Bank Claim	Consolidated claims of a banking sector towards all sectors in a non-Eurozone country (Immediate Counterparty Basis). Data gaps are linearly interpolated.	BIS Consolidated Banking Statistics
Weight	Share of Bank Claims towards a non-Eurozone country compared to total Bank Claims.	
Stock Market Returns	Daily change in natural logarithm of (market-wide) stock market returns of a non-Eurozone country in US Dollar.	Datastream, FTSE, MSCI, S&P
Exposure-Weighted Returns	Weight towards non-Eurozone country times StockMarketReturn of the same non-Eurozone country.	
NonEZStockReturns	Sum of ExposureWeightedReturns of every Eurozone country in the sample.	

Daily Financial Market Data (all return variables winsorized at 1st and 99th percentile)		
Sovereign Distress	Daily change in the natural logarithm of the benchmark 10-year Datastream government bond index of Germany minus the corresponding index change of a sample country.	Datastream
VIX	Daily change in VIX volatility index.	CBOE
US Corporate Credit Spread	Daily change in spread between the Thomson Reuters corporate benchmark BBB 10-years yield and the corresponding AAA 10-years yield.	Thomson Reuters
US Term Spread	Daily change in spread between 10-years US Treasury yield and the 3-Months US T-Bill yield.	Datastream, Federal Reserve
VSTOXX	Daily change in Vstoxx volatility index.	STOXX
Nominal Euro Exchange Rate	Daily change of natural logarithm of nominal effective exchange rate of the Euro.	JP Morgan
Non-Financial Itraxx	The daily change of the natural log of the 10-year Itraxx Europe is regressed against the daily change of 10-Year Itraxx senior and subordinated financial indices. The residuals from this regression are the non-financial Itraxx.	Markit
Current Account Holdings	Daily change in the natural log of current account holdings (sum of minimum and excess reserves by banks at the ECB).	ECB
Eurozone Term Spread	Daily change in the spread between EuroMTS Government 7-10 year broad yield and the 3-month Euribor yield.	European Banking Federation, FTSE MTS
Non-Financial Stock Market Returns	Daily change in the natural logarithm of the non-financial stock indices is orthogonalized towards NationalBankDistress and daily changes in the Eurozone total stock market returns.	Datastream

Variables used in Robustness Section (all return variables winsorized at 1st and 99th percentile)		

Sovereign Bond Yield Spread	Daily change in the yield spread between a country's benchmark 10-year government bond and the corresponding German bond.	Datastream
CDS Spread	Daily change in CDS spread of a 5-year US Dollar CDS rate of a country and the corresponding German CDS rate.	Thomson Reuters
Bank Distress	BankReturns times -1.	
NationalBankDistress by Orthogonalization	BankReturns orthogonalized to EurozoneBankReturns. Variable is then multiplied times -1.	
NonEZBankStock Returns	NonEZStockReturns using only bank stocks.	Datastream
NonEZStockReturns Different elimination of EZ component	NonEZStockReturns is orthogonalized towards the first principal component of BankReturns.	
NonEZStockReturns Excluding EU borrowers	NonEZStockReturns excluding EU countries without the Euro from exposure portfolio.	
NonEZStockReturns Excluding large credit suppliers	NonEZStockReturns excluding non-EZ countries that hold claims equal or beyond 90th percentile towards an EZ country from portfolio.	
Trade-Weighted Shocks	Export weight of EZ towards non-EZ country times non-financial stock market returns. Variable is orthogonalized towards NonEZStockReturns.	IMF (Direction of Trade), Datastream
Non-Financial Stock Market Returns	Daily change in natural logarithm of non-financial stock returns of non-Eurozone country in local currency.	Datastream
Non-Financial Stock Market Returns: Alt. Version	The daily change in the natural logarithm of the non-financial stock indices of every country is orthogonalized towards daily changes in the Eurozone total stock market returns.	Datastream
Euro-Dollar ER	Daily change in the natural log of the Euro to Dollar rate.	Thomson Reuters
Trading Liquidity	Ratio of each bank's stock turnover to stocks outstanding; Then asset-weighted and aggregated on country level.	Datastream

Interaction Variables applied in appendix chapter (all variables trimmed at 1st and 99th percentile)

Debt to GDP	General government consolidated gross debt in % of GDP.	Eurostat
GDP Growth	Quarterly GDP growth in market prices (quarterly).	OECD
Unemployment Rate	Harmonised unemployed rates, all persons, all ages, seasonally adjusted (monthly).	OECD
Fiscal-Deficit-to-GDP	Net lending (-) / net borrowing (+) of general government in % of gross domestic product (quarterly); Oxford Economics data used for all countries except Netherlands and Portugal, which uses Eurostat data, as Oxford data is noisy for these countries.	Oxford Economics, Eurostat
Current Account to GDP	Current Account to GDP Ratio (quarterly).	OECD
Inflation	Annual rate of change of harmonised index of consumer prices (monthly).	Eurostat
Government Bond Issuance to GDP	Gross issuance of debt securities of general government in % of GDP (monthly). The data (and all other issuance/redemption data) starts only in December 2009.	ECB
Government Bond Redemption to GDP	Gross redemption of debt securities of general government in % of GDP (monthly).	ECB

Variable	Description	Source
High Issuance	Dummy variable equal to one in months in which the gross issuance of government debt securities was above the sample median of this country.	ECB
High Redemption	Dummy variable equal to one in months in which the gross redemption of government debt securities was above the sample median of this country.	ECB
Home Share	Share of domestic government bonds held by a banking sector in relation to total government bonds held.	ECB
Non-Performing Loans Ratio	Non-Performing loans to total gross loans ratio (quarterly). Adjustment in case of reporting gaps: If gap is two periods or shorter: gap is interpolated. If gap is more than two periods: gap is replaced with yearly value.	IMF Financial Soundness Indicators
Return on Assets	Return on assets ratio of banking sector (quarterly). Same adjustment as for Non-Performing Loans Ratio.	IMF Financial Soundness Indicators
Tier1 Capital Ratio	Regulatory Tier1 capital to risk-weighted asset ratio. Same adjustment as for Non-Performing Loans Ratio.	IMF Financial Soundness Indicators
Bank Asset to Public Revenue Ratio	Ratio of monthly total balance sheet size of banking sector towards total general government revenue. The latter uses the same data sources as Fiscal-Deficit-to-GDP (monthly).	ECB, Oxford Economics, Eurostat
Central Bank Funding Share	Share of central bank funding in bank's liabilities (monthly).	ECB
Macroprudential Regulation Index	Index aggregating the number of implemented macroprudential instruments (quarterly). Data is only available until 2014:Q4.	Cerutti et al. (2016)
Non-Financial Securities to GDP	Monthly outstanding securities issued by non-financial corporations in ratio to GDP (monthly).	ECB
Policy Uncertainty Europe	Economic policy uncertainty index for Europe (monthly).	policy uncertainty.com Baker et al. (2016)
Policy Uncertainty Euro	First component of principal component analysis of the policy uncertainty indices of Ireland, Spain and Italy (monthly).	policy uncertainty.com Baker et al. (2016)
Election	Dummy that is 1 if the country held a parliamentary election in that month.	ParlGov Database
Left/Right Government	Weighted index of left/right party preference of cabinet coalition. Preferences of ruling parties are weighted based on their seats in parliament. Higher index means more conservative government (monthly).	ParlGov Database
State/Market Government	Weighted index of state/market-friendly party preference of cabinet coalition. Preferences of ruling parties are weighted based on their seats in parliament. Higher index means more market-friendly government (monthly).	ParlGov Database
Contra/Pro-EU Government	Weighted index of contra/pro-EU party preference of cabinet coalition. Preferences of ruling parties are weighted based on their seats in parliament. Higher index means more EU-friendly government (monthly).	ParlGov Database

References to Chapter 2

Acharya, V., Drechsler, I. & Schnabl, P. (2014), 'A pyrrhic victory? Bank bailouts and sovereign credit risk', *The Journal of Finance* **69**(6), 2689–2739.

Acharya, V. V., Eisert, T., Eufinger, C. & Hirsch, C. (2018), 'Real effects of the sovereign debt crisis in Europe: Evidence from syndicated loans', *The Review of Financial Studies* **31**(8), 2855–2896.

Acharya, V. V., Pierret, D. & Steffen, S. (2018), 'Lender of last resort versus buyer of last resort–evidence from the European sovereign debt crisis'. ZEW - Centre for European Economic Research Discussion Paper No. 16-019.

Acharya, V. V. & Steffen, S. (2015), 'The "greatest" carry trade ever? Understanding Eurozone bank risks', *Journal of Financial Economics* **115**(2), 215–236.

Aizenman, J., Hutchison, M. & Jinjarak, Y. (2013), 'What is the risk of European sovereign debt defaults? Fiscal space, CDS spreads and market pricing of risk', *Journal of International Money and Finance* **34**, 37–59.

Altavilla, C., Pagano, M. & Simonelli, S. (2017), 'Bank exposures and sovereign stress transmission', *Review of Finance* **21**(6), 2103–2139.

Alter, A. & Schüler, Y. S. (2012), 'Credit spread interdependencies of European states and banks during the financial crisis', *Journal of Banking & Finance* **36**(12), 3444–3468.

Baker, S. R., Bloom, N. & Davis, S. J. (2016), 'Measuring economic policy uncertainty', *The Quarterly Journal of Economics* **131**(4), 1593–1636.

Battistini, N., Pagano, M. & Simonelli, S. (2014), 'Systemic risk, sovereign yields and bank exposures in the Euro crisis', *Economic Policy* **29**(78), 203–251.

Beck, T., De Jonghe, O. & Mulier, K. (2017), 'Bank sectoral concentration and (systemic) risk: Evidence from a worldwide sample of banks'. CEPR Discussion Paper No. DP12009.

Becker, B. & Ivashina, V. (2017), 'Financial repression in the European sovereign debt crisis', *Review of Finance* **22**(1), 83–115.

Beirne, J. & Fratzscher, M. (2013), 'The pricing of sovereign risk and contagion during the European sovereign debt crisis', *Journal of International Money and Finance* **34**, 60–82.

Beltratti, A. & Stulz, R. M. (2012), 'The credit crisis around the globe: Why did some banks perform better?', *Journal of Financial Economics* **105**(1), 1–17.

Bongini, P., Laeven, L. & Majnoni, G. (2002), 'How good is the market at assessing bank fragility? A horse race between different indicators', *Journal of Banking & Finance* **26**, 1011–1028.

Breckenfelder, J. & Schwaab, B. (2018), 'Bank to sovereign risk spillovers across borders: Evidence from the ECB's comprehensive assessment', *Journal of Empirical Finance* **49**, 247–262.

Brunnermeier, M. K., Garicano, L., Lane, P. R., Pagano, M., Reis, R., Santos, T., Thesmar, D., Van Nieuwerburgh, S. & Vayanos, D. (2016), 'The sovereign-bank diabolic loop and ESBies', *The American Economic Review Papers and Proceedings* **106**(5), 508–512.

Buch, C. M., Koetter, M. & Ohls, J. (2016), 'Banks and sovereign risk: A granular view', *Journal of Financial Stability* **25**, 1–15.

Buch, C. M. & Neugebauer, K. (2011), 'Bank-specific shocks and the real economy', *Journal of Banking & Finance* **35**(8), 2179–2187.

Bun, M. J. & Harrison, T. D. (2019), 'OLS and IV estimation of regression models including endogenous interaction terms', *Econometric Reviews* **38**(7), 814–827.

Cameron, A. C., Gelbach, J. B. & Miller, D. L. (2008), 'Bootstrap-based improvements for inference with clustered errors', *The Review of Economics and Statistics* **90**(3), 414–427.

Cerutti, M. E. M., Correa, M. R., Fiorentino, E. & Segalla, E. (2016), 'Changes in prudential policy instruments - a new cross-country database'. IMF Working Paper, WP/16/110.

CNBC (2010), '"no bailout" clause? The EU's Greek rescue problems'. https://www.cnbc.com/id/35327584.

CNN (2016), 'Italy: Europe's next domino to fall?'. https://money.cnn.com/2016/07/06/investing/banks-italy-brexit-crisis/index.html.

Crosignani, M. (2017), 'Why are banks not recapitalized during crises?'. FEDS Working Paper No. 2017-084.

Crosignani, M., Faria-e Castro, M. & Fonseca, L. (2020), 'The (unintended?) consequences of the largest liquidity injection ever', *Journal of Monetary Economics* **112**, 97–112.

De Bruyckere, V., Gerhardt, M., Schepens, G. & Vander Vennet, R. (2013), 'Bank/sovereign risk spillovers in the European debt crisis', *Journal of Banking & Finance* **37**(12), 4793–4809.

De Grauwe, P. & Ji, Y. (2013), 'Self-fulfilling crises in the Eurozone: An empirical test', *Journal of International Money and Finance* **34**, 15–36.

De Marco, F. & Macchiavelli, M. (2016), 'The political origin of home bias: The case of Europe'. FEDS Working Paper No. 2016-060.

Demirgüç-Kunt, A. & Huizinga, H. (2013), 'Are banks too big to fail or too big to save? International evidence from equity prices and CDS spreads', *Journal of Banking & Finance* **37**(3), 875–894.

Drechsler, I., Drechsel, T., Marques-Ibanez, D. & Schnabl, P. (2016), 'Who borrows from the lender of last resort?', *The Journal of Finance* **71**(5), 1933–1974.

Eichler, S. & Sobański, K. (2016), 'National politics and bank default risk in the Eurozone', *Journal of Financial Stability* **26**, 247–256.

Fahlenbrach, R. & Stulz, R. M. (2011), 'Bank CEO incentives and the credit crisis', *Journal of financial economics* **99**(1), 11–26.

Farhi, E. & Tirole, J. (2017), 'Deadly embrace: Sovereign and financial balance sheets doom loops', *The Review of Economic Studies* **85**(3), 1781–1823.

Financial Times (2011), 'EU leaders agree €109bn Greek bail-out'. `https://www.ft.com/content/952e0326-b3af-11e0-855b-00144feabdc0`.

Financial Times (2012), 'Spain takes 45% stake in Bankia'. `https://www.ft.com/content/29595d88-99e8-11e1-accb-00144feabdc0`.

Financial Times (2016), 'Monte Paschi struggles most in EU bank stress tests'. `https://www.ft.com/content/b1c93b2d-9ccb-3faf-a4ae-7b676fe7c015`.

Fratzscher, M. & Rieth, M. (2019), 'Monetary policy, bank bailouts and the sovereign-bank risk nexus in the Euro Area', *Review of Finance* **23**(4), 745–775.

Gaballo, G. & Zetlin-Jones, A. (2016), 'Bailouts, moral hazard and banks' home bias for sovereign debt', *Journal of Monetary Economics* **81**, 70–85.

Gerlach, S., Schulz, A. & Wolff, G. B. (2010), 'Banking and sovereign risk in the Euro Area'. Bundesbank Discussion Paper Series 1 No. 09/2010.

Gropp, R., Vesala, J. & Vulpes, G. (2006), 'Equity and bond market signals as leading indicators of bank fragility', *Journal of Money, Credit and Banking* pp. 399–428.

Horváth, B. L., Huizinga, H. & Ioannidou, V. (2015), 'Determinants and valuation effects of the home bias in European banks' sovereign debt portfolios'. CEPR Discussion Paper No. DP10661.

IMF (2010), 'IMF approves €22.5 billion loan for Ireland'. `https://www.imf.org/en/News/Articles/2015/09/28/04/53/socar121610a`.

Kallestrup, R., Lando, D. & Murgoci, A. (2016), 'Financial sector linkages and the dynamics of bank and sovereign credit spreads', *Journal of Empirical Finance* **38**, 374–393.

Kirschenmann, K., Korte, J. & Steffen, S. (2017), 'The zero risk fallacy-banks' sovereign exposure and sovereign risk spillovers'. ZEW Discussion Papers, No. 17-069.

Longstaff, F. A., Pan, J., Pedersen, L. H. & Singleton, K. J. (2011), 'How sovereign is sovereign credit risk?', *American Economic Journal: Macroeconomics* **3**(2), 75–103.

Moody's (2010), 'Moody's downgrades Irish banks further to sovereign downgrade'. `https://www.moodys.com/research/Moodys-downgrades-Irish-Banks-further-to-sovereign-downgrade--PR_211346`.

Moody's (2015), 'Moody's downgrades Greek banks' senior unsecured debt ratings to C and confirms deposit ratings at Caa3'. `https://www.moodys.com/research/Moodys-downgrades-Greek-banks-senior-unsecured-debt-ratings-to-C--PR_333800`.

New York Times (2010), 'Ireland unveils austerity plan to help secure bailout'. `https://www.nytimes.com/2010/11/25/world/europe/25ireland.html`.

New York Times (2012), 'Spain to accept rescue from Europe for its ailing banks'. https://www.nytimes.com/2012/06/10/business/global/spain-moves-closer-to-bailout-of-banks.html.

New York Times (2013), 'Real estate losses weigh on Santander'. https://dealbook.nytimes.com/2013/01/31/santanders-profit-hit-by-real-estate-concerns/.

Nizalova, O. Y. & Murtazashvili, I. (2016), 'Exogenous treatment and endogenous factors: Vanishing of omitted variable bias on the interaction term', *Journal of Econometric Methods* **5**(1), 71–77.

Ongena, S., Popov, A. & Van Horen, N. (2019), 'The invisible hand of the government: Moral suasion during the European sovereign debt crisis', *American Economic Journal: Macroeconomics* **11**(4), 346–79.

Popov, A. & Van Horen, N. (2015), 'Exporting sovereign stress: Evidence from syndicated bank lending during the Euro Area sovereign debt crisis', *Review of Finance* **19**(5), 1825–1866.

Reuters (2010), 'S&P downgrades Greece ratings into junk status'. https://www.reuters.com/article/greece-ratings-sandp/sp-downgrades-greece-ratings-into-junk-status-idUSWNA964520100427.

Reuters (2012), 'Spain beset by bank crisis, downgrades, bond pressure'. https://www.reuters.com/article/us-spain-economy-idUSBRE84G0CK20120517?feedType=RSSfeedName=topNewsutm_source=feedburnerutm_medium=feedutm_campaign=Feed:+reuters/topNews+(News+/+US+/+Top+News)utm_content=Google+International.

Reuters (2014), 'Portugal in $6.6 billion rescue of Banco Espirito Santo'. https://www.reuters.com/article/us-portugal-bes-cenbank/portugal-in-6-6-billion-rescue-of-banco-espirito-santo-idUSKBN0G30T620140804.

Roodman, D., Nielsen, M. Ø., MacKinnon, J. G. & Webb, M. D. (2019), 'Fast and wild: Bootstrap inference in Stata using boottest', *The Stata Journal* **19**(1), 4–60.

Schnabel, I. & Schüwer, U. (2016), 'What drives the relationship between bank and sovereign credit risk?'. Beitraege zur Jahrestagung des Vereins fuer Socialpolitik 2017: Alternative Geld- und Finanzarchitekturen.

Singh, M. K., Gómez-Puig, M. & Sosvilla-Rivero, S. (2016), 'Sovereign-bank linkages: Quantifying directional intensity of risk transfers in EMU countries', *Journal of International Money and Finance* **63**, 137–164.

The Guardian (2010a), 'Angela Merkel dashes Greek hopes of rescue bid'. https://www.theguardian.com/theguardian/2010/feb/11/germany-greece-merkel-bailout-euro.

The Guardian (2010b), 'Anglo Irish Bank downgrade raises pressure on Ireland'. https://www.theguardian.com/business/2010/sep/27/anglo-irish-downgrade.

The Guardian (2010c), 'EU debt crisis: Greece granted €110bn aid to avert meltdown'. https://www.theguardian.com/world/2010/may/02/eu-debt-crisis-greece-aid-meltdown.

The Guardian (2010*d*), 'EU ministers agree Greek bailout terms'. `https://www.theguardian.com/world/2010/apr/11/eu-greece-bailout-terms`.

The Guardian (2010*e*), 'Ireland bailout: Full Irish government statement'. `https://www.theguardian.com/business/ireland-business-blog-with-lisa-ocarroll/2010/nov/28/ireland-bailout-full-government-statement`.

The Guardian (2011*a*), 'Portugal bailout details boost euro and bond markets'. `https://www.theguardian.com/business/2011/may/04/portugal-bailout-euro-rises-bond-markets`.

The Guardian (2011*b*), 'Portugal's credit rating downgraded to junk status'. `https://www.theguardian.com/world/2011/jul/06/portugal-credit-rating-junk-status`.

The Guardian (2012), 'Eurozone bank bailout deal throws lifeline to Spain and Italy'. `https://www.theguardian.com/business/2012/jun/29/eurozone-bank-bailout-spain-italy`.

The Guardian (2016), 'Quiet crisis: Why battle to prop up Italy's banks is vital to EU stability'. `https://www.theguardian.com/world/2016/may/10/battle-prop-up-italy-banks-eu-brexit-grexit-bad-loans`.

Wooldridge, J. M. (2010), *Econometric analysis of cross section and panel data*, MIT press.

Chapter 3:

What drives the Commodity-Sovereign Risk Dependence in Emerging Market Economies?

3.1 Introduction

Global commodity price cycles have been among the most influential drivers of sovereign defaults in history (Reinhart et al. 2016). Higher export commodity prices improve sovereign solvency by spurring economic growth and tax revenues, by increasing the profitability of state-owned commodity enterprises and by generating inflows of foreign exchange thus increasing the government's ability to service its external debt. Fluctuations in commodity prices are therefore important business cycle drivers, in particular for emerging market economies (Fernández et al. (2018), Fernández et al. (2017)), while also comprising political considerations: In the upturn of the commodity cycle, especially autocratic regimes with poor institutions tend to build up unsustainable debt levels, which can lead to debt overhangs and default in the downturn of the cycle (Arezki & Brückner 2012).

Commodity cycles matter, as during the 2013-2017 period, 102 out of 189 countries in the world were considered to be commodity-dependent according to UNCTAD (2019). Both the literature and policy reports often make commodity dependence responsible for creating vulnerabilities: Globally determined raw material prices steer the economic performance and the costs to borrow money on financial markets of commodity-dependent countries beyond their control.[21] Despite its relevance, a comprehensive study of the economic drivers of commodity-sovereign risk dependence is lacking in the literature.

We contribute to the literature by analyzing the magnitude and determinants of commodity-sovereign risk dependence from the viewpoint of daily financial market investors. To this end, we build a daily panel of 34 emerging market economies from January 1, 1994 to December 31, 2016. We measure commodity-sovereign risk dependence as the relationship between a country's sovereign creditworthiness (measured by changes in the Emerging Market Bond Index (EMBI) yield spread relative to US Treasuries) and the returns of its export-weighted commodity price index. We control thoroughly for global developments on financial markets and most importantly for a country's general stock return which should account for major economic movements each day. Any impact of commodity prices on sovereign creditworthiness beyond these controls is likely due to raw material prices affecting the fiscal situation, investment possibilities and general economic outlook of a country and hence imply commodity dependence.

Our empirical approach tests the hypothesis if and by how much export-weighted commodity returns affect sovereign risk. We find that an increase in our commodity performance measure by one standard deviation is associated on average with a reduction of the EMBI yield spread of 33 basis points (bps), which corresponds to 3.5% of the EMBI spread's standard deviation and indicates a decrease in sovereign risk. For countries with a commodity export share on total exports equal or above the 90th percentile,[22] a one standard deviation increase in export commodity prices is associated with a 47.5 bps decrease in the EMBI spread (corresponding to 5% of the EMBI's standard deviation). Although we can only explain a modest fraction of bond price changes in absolute terms, the standardized effect of commodity prices on EMBI bond yield spreads are around 40% as large as for the VIX or the U.S. corporate bond spread and thus relatively important in pricing emerging market bonds.

---

[21] An article by the World Economic Forum (2019) from 17 May 2019 entitled *"We must help developing countries escape commodity dependence"* says: *"When a country's economy is not diversified and relies heavily on basic products, it puts itself at the mercy of international market prices. When prices go down, employment, exports and government revenue suffer. (...) [P]utting too many eggs in one basket renders the country vulnerable."*

[22] Affected countries are Ecuador, Ghana, Kazakhstan, Nigeria, Peru and Venezuela.

Our second contribution focuses on the heterogeneous nature of commodity-sovereign risk-dependence. To the best of our knowledge, we are the first to examine a broad set of hypotheses on possible conditioning factors shaping the size of this dependence such as a country's commodity exporting structure, its macroeconomic conditions, and the implementation of policy measures that might reduce commodity dependence. We find strong heterogeneous effects that differentiate the average magnitude of commodity dependence reported above.

Looking at the structure of the country's commodity export industry, we find that countries with greater commodity exports on total export shares are significantly more commodity-dependent (pointing to a higher relevance of commodity price fluctuations for economic growth, tax revenues or direct cash flows from the commodity sector due to public stakes or royalties). When analyzing specializations into different commodity subgroups, we observe that countries which export predominantly energy face larger, while countries focusing on industrial metals are subject to smaller commodity dependence. We do not find significant evidence that the volatility of exported commodities and the degree of diversification of commodity exports affect the commodity-sovereign risk dependence.

When analyzing the impact of macroeconomic factors on commodity-sovereign risk dependence, we find that the reliance on commodities for sovereign funding increases in economic recessions (lower GDP growth) and, likely associated therewith, when public or private sectors lack fiscal resources (lower tax revenues or corporate profits). We find no statistically significant connection between commodity dependence and GDP per capita or the inflation rate. A surprising coefficient is the interaction of commodity performance and public debt, which suggests higher commodity dependence at lower debt levels, however, the coefficient is only weakly significant.

We find only weak evidence that commodity dependence increases in times of sovereign debt crises. However, we uncover that the repayment history of a sovereign matters in that countries with a very distant or no incidents of sovereign default display lower commodity dependence.

Beyond national macroeconomic factors, we find that commodity dependence of emerging markets increases significantly in times of more expansionary U.S. monetary policy. This observation is in line with findings in the literature that an accommodative U.S. monetary weakens the dollar, which tends to boost commodity prices (Akram 2009). Moreover, U.S.

interest rates affect demand and supply conditions in commodity markets (Frankel 2006) and lower U.S. interest rates lead to lower risk aversion, rising capital flows and foreign lending activities into emerging markets (Bräuning & Ivashina 2020, Ahmed & Zlate 2014, Temesvary et al. 2018).

When turning to policy measures that are likely able to mitigate commodity dependence, we find clear support that countries with stronger institutions are significantly less reliant on the price performance of their main commodity exports for their sovereign creditworthiness. Improving institutional quality, measured with control of corruption, rule of law, political stability or more progressive tax systems, likely makes sure that clear ownership rights for extracted raw materials exist, that rent extraction is limited or that gains from commodity exports are efficiently distributed.[23] Some papers stress that larger endowments of natural resources make it more difficult to improve institutional quality (Arezki & Brückner 2011a, Gylfason 2001). We take this issue into account by limiting our estimation to those countries that are heavy commodity exporters. Our main results continue to hold, suggesting that even among strong commodity exporters, those with better institutions seem to fare better in terms of lower commodity dependence, which is also in line with Bhattacharyya & Hodler (2010), Mehlum et al. (2006) and Arezki & Brückner (2011b).

Further measures that are associated with alleviating commodity dependence are attracting more FDI inflows, investing in physical capital and infrastructure, and building larger manufacturing sectors. We conclude from these results that fostering downstream production technologies and diversifying economic activities can be successful ways for countries to reduce commodity dependence. On the other hand, we find only limited evidence that speaks in favor of mitigating commodity dependence by means of development assistance or loans.

Countries that build up higher reserve assets in relation to GDP face significantly lower commodity dependence by reducing their reliance on foreign exchange inflows via commodity exports. This result suggests that low official reserve buffers make emerging markets particularly vulnerable to international commodity price fluctuations. Lastly, we find that countries

---

[23]For instance, Frankel (2010) discusses the successful fiscal rule of Chile that is also mentioned in an article in The Economist (2017) from 5th October 2017 entitled *"Commodities are not always bad for you"*: *"Resource-rich economies need equally resourceful macroeconomic policies. One of the best examples is Chile. Its fiscal rule curbs government spending when the copper price exceeds its long-term trend, as judged by an independent committee of experts. During good times, fiscal restraint makes room for mining to boom without unduly squeezing the rest of the economy. During bad times, it leaves scope for fiscal easing to offset the damage."*

with shielded trade and financial accounts are subject to a significantly stronger commodity dependence.

Our work builds on seminal papers in the literature such as Deaton (1999), Sachs & Warner (1995) and Sachs & Warner (2001) that highlight the tight connection between GDP growth and commodity prices of raw material exporting countries, and which is also shown in more recent work by Drechsel & Tenreyro (2018) and Fernández et al. (2017). Several papers discuss the implications of the "resource curse" of developing countries (see Frankel (2010) for an overview) which is, however, disputed by other authors (James 2015, Alexeev & Conrad 2009, Davis 1995).

Several papers study the relevance of commodity prices for determining sovereign risk. Arezki & Brückner (2011b) study the effect of windfall gains from commodity price shocks on sovereign bond yield spreads. They find that higher commodity prices reduce sovereign yield spreads in democratic regimes and increase yield spreads in autocratic regimes, pointing to the resource curse in countries with poor institutions. Similar results are found for countries' external debt ratios in Arezki & Brückner (2012). Hilscher & Nosbusch (2010) use export commodity prices to instrument terms of trade and their effect on sovereign bond spreads at an annual frequency. They find that the level and volatility of terms of trade explain a huge fraction of annual variation of sovereign yield spreads. Aizenman et al. (2016) find that higher volatility of commodity terms of trade is associated with an increase in sovereign CDS.[24]

We contribute to this literature by studying the channels shaping the commodity dependence of sovereign default risk (such as the size, volatility or diversification of the country's commodity exporting regime, the stance of the domestic economy, monetary policy, capital controls, as well as institutional and policy factors). Thereby, we aim at explaining the heterogeneity of the effects of commodity price shocks to sovereign solvency across different emerging markets which informs the policy debate on how to curb the commodity-sovereign risk-nexus. Furthermore, we use daily data instead of quarterly or yearly averages. Daily variation in sovereign bond and commodity prices is less susceptible to endogenously formed policy decisions: On a yearly basis, policy makers might adjust e.g. institutions with respect

---

[24]A related literature shows that commodity prices determine the value of commodity currencies, by shaping terms of trade and the generation of foreign exchange revenues (Chen & Rogoff (2003); Cashin et al. (2004); Kohlscheen et al. (2017)).

to longer-term commodity price changes. On a daily basis, institutional quality is given and cannot be adjusted to cushion, for instance, the impact of a negative commodity price shock hitting a country on this day.

The rest of this article is organized as follows: In Section 3.2, we describe the data we use in order to isolate commodity dependence. Section 3.3 presents our empirical strategy and reports our baseline results. Following on this, Section 3.4 investigates the drivers of commodity dependence and discusses the effect of policy measures to tackle it. We conduct encompassing robustness checks in Section 3.5. Section 3.6 concludes.

3.2 Data, Variables and Summary Statistics

3.2.1 Dependent Variable: Sovereign Default Risk

Our sovereign default risk measure is drawn from the Emerging Market Bond Index (EMBI) provided by J.P. Morgan. Sovereign bonds that are issued by emerging markets and included in the EMBI are US dollar-denominated which rules out exchange rate risk. Issued debt must furthermore have more than one year to maturity and exceed an outstanding face value of $500 million to be eligible for the EMBI. For these reasons, EMBI returns are a well standardized, widely used and liquid measure to track the daily performance of emerging market sovereign debt. For our analysis, we use EMBI Global data as it covers more instruments than the original EMBI+ index and has better data availability.

The introduction of the EMBI Global at the end of 1993 determines the beginning of our estimation period which is set from January 1, 1994 to December 31, 2016, though some countries enter only at later points in time. We collect EMBI Global data for a panel of 34 countries which can be found in Table 3.1. Though more countries with EMBI data exist, data availability with respect to other variables, in particular stock returns, restricts our sample to the set of the countries listed below. To make sure every country included has sufficient variation, we include a country if it has liquid EMBI data for at least nine years, i.e. at least since 2008.[25]

---

[25]Though we could also choose a ten year inclusion rule, the countries Ghana, Jamaica, Kazakhstan and Sri Lanka start reporting EMBI data in 2007. Also, Thailand reports a nine year EMBI period from 1997 to 2006. To include these countries, we set the threshold at nine years.

Table 3.1: Summary statistics by country

Region	Country	Mean ΔEMBI	Mean ΔCommodity-Performance	Number of ΔEMBI and ΔCommodity-Performance observations in baseline regression
Africa	Egypt	-0.0215	0.00292	3910
	Ghana	-0.0967	0.0125	1665
	Ivory Coast	0.839	0.00503	1937
	Morocco	-0.187	0.0177	3349
	Nigeria	-0.240	0.0142	3658
	South Africa	-0.0733	0.00888	4149
	Tunisia	-.00262	0.0224	2255
Americas	Argentina	0.316	0.0108	5975
	Brazil	-0.405	0.00502	5833
	Chile	-0.0173	0.0181	4519
	Colombia	-0.145	0.0135	5142
	Ecuador	-0.136	0.0263	4746
	Jamaica	0.0173	0.0212	2186
	Mexico	-0.133	0.0217	5971
	Panama	-0.118	0.0135	2137
	Peru	-0.169	0.0103	5116
	Venezuela	-0.0985	0.0393	4826
Asia	China	-0.0126	0.00739	5796
	Indonesia	-0.139	0.0187	3225
	Kazakhstan	-0.109	0.0128	2436
	Malaysia	-0.0250	0.0219	5131
	Pakistan	-0.192	-0.000387	3243
	Philippines	-0.0937	0.00231	4889
	Russia	-0.125	0.0155	4928
	Sri Lanka	-0.0458	0.0236	2328
	Thailand	-0.153	0.0183	2209
	Turkey	-0.185	0.00126	5315
	Vietnam	-0.0810	0.0137	2573
Europe	Bulgaria	-0.306	0.0109	2229
	Croatia	-0.169	.0153	2374
	Hungary	-0.0385	-0.00276	4641
	Poland	-0.0748	0.0114	5681
	Serbia	-0.0909	0.0102	2229
	Ukraine	-0.171	0.00612	2715

Sample period is 01/01/1994–31/12/2016. ΔEMBI (first difference of natural log of Emerging-Market-Bond-Index) is winsorized at 5th and 95th, ΔCommodityPerformance (export-weighted commodity returns) is winsorized at 1st and 99th percentile. See Table A.3.3 for definitions and sources.

Data is drawn on a daily frequency to exploit maximum data variation and give our estimation strategy the perspective of market participants that incorporate daily news into their investment behavior. Our dependent variable is the daily first difference of a country's EMBI yield spread (relative to the U.S. treasury rate), so that positive changes in the EMBI spread indicate declining sovereign creditworthiness (or increasing sovereign risk).

While all other data is winsorized at the 1st and 99th percentile, we winsorize EMBI spread data on the 5th and 95th percentile because the raw spread differences have occasionally extreme values. We account for episodes with temporarily illiquid country EMBI indices by dropping observations with zero changes in the EMBI spread that occur for more than two consecutive trading days. In a robustness check, we also drop all countries exhibiting such periods of low liquidity and find results in line with our main specification.

3.2.2 Deriving Country-specific Commodity Performance

We construct the daily export-weighted commodity performance by weighting commodity price returns with the country's commodity export shares. In order to determine which commodities are to be included in the export portfolio of each country, we refer to the commodities comprising the Goldman Sachs Commodity Index (GSCI) provided by S&P. The GSCI provides daily spot index data of 24 commodities in the main index. Each commodity can be grouped either under agriculture, livestock, industrial metals, precious metals or energy. Each of these sub-groups also has its own aggregated group price index. The GSCI includes commodity types based on global production values and the availability of active and liquid futures markets. Commodities in the index are therefore frequently traded and priced in U.S. dollar which is in contrast to many regional commodity price data sources that often suffer from periods of poor liquidity. By using GSCI data, we make sure that our commodity portfolio measures include both highly relevant and globally-priced commodities. Table 3.2 contains a list of all commodities.

We match commodity price to commodity export data derived from the UN's Comtrade Database and ITC's Trade Map. Most commodity export volumes can be directly matched to their corresponding prices. However, some price series start after the beginning of our sample period in 1994 or have only a roughly corresponding export match. This issue concerns energy and petroleum-based commodity prices which are included in the GSCI as WTI crude oil, Brent crude oil, gas oil, heating oil, gasoline and natural gas. As there is no perfect export match for all of these commodities, for instance if a crude oil export is classified under WTI or Brent standards, and because price data for Brent crude oil and gas oil starts only after 1994, we aggregate these commodities under their sub-group price index, i.e. energy. The matching export data includes all crude oil and petroleum gas exports. Since all price returns within the energy group are highly correlated, we believe that this sub-group-level aggregation is the most precise way to capture and price petroleum-based exports and to avoid a potentially biasing match between not fully overlapping price and export data.

We further aggregate the GSCI price series Kansas wheat and CBOT wheat under the aggregated price series "All Wheat" and the series for feeder cattle and live cattle under the aggregated spot index of "All Cattle". Table 3.2 reports the final match between commodity price and export data.

Export volumes for different commodities are available on a yearly (y) frequency only. We therefore calculate the share of each commodity on the total commodity exports of each country as a yearly-varying weighting factor.

Each daily commodity price return is then multiplied by its country-specific weighting factor. However, we lag the export weights by one year in order to rule out commodity prices mechanically affecting commodity export weights. In a robustness check in Section 3.5, we also report results with constant commodity weights observed at the beginning of the sample, which leads to robust conclusions. We aggregate the weighted commodity returns over all commodities c on a daily basis t for each country i, arriving at a country-specific commodity return measure in which the largest commodity exports have the greatest weight:

$$CommodityPerformance_{it} = \sum_{c} CommodityExportShare_{icy-1} * \Delta CommodityPrice_{ct}$$

We will also test different versions of the commodity performance index to control for world-market relevant exporters, a dummy-version, and net- instead of gross-exports in the alternative specification section.

Table 3.2: Match between commodity prices and export quantities

GSCI single commodity index	GSCI sub-group index	GSCI group index	Matching commodity export
Cocoa			1801: Cocoa beans
Coffee			090111: Coffee (excluding roasted and decaffeinated)
Corn			1005: Maize or corn
Cotton		Agri-culture	52: Cotton
Soybeans			1201: Soya beans, whether or not broken
Sugar			1701: Cane or beet sugar and chemically pure sucrose
Wheat (CBOT)	**All**		1001: Wheat and meslin
Wheat (Kansas)	**wheat**		
Lean Hogs			0103: Live swine
Feeder Cattle	**All**	Livestock	010229: Live cattle
Live Cattle	**cattle**		
Brent Crude Oil			2709: Petroleum oils and oils obtained from bituminous minerals, crude
WTI Crude Oil			
Gas Oil		**Energy**	
Heating Oil			2711: Petroleum gas and other gaseous hydrocarbons
RBOB Gasoline			
Natural Gas			
Aluminum			2606: Aluminium ores and concentrates 7601: Unwrought aluminium
Copper			2603: Copper ores and concentrates 7402: Copper, unrefined
Lead		Industrial Metals	2607: Lead ores and concentrates 7801: Unwrought lead
Nickel			2604: Nickel ores and concentrates 7502: Unwrought nickel
Zinc			2608: Zinc ores and concentrates 7901: Unwrought zinc
Gold			7108: Gold, (...) unwrought or not further worked than semi-manufactured or in powder form
Silver		Precious Metals	261610: Silver ores and concentrates 7106: Silver, (...) unwrought or in semi-manufactured forms, or in powder form

Match between commodity prices (from GSCI) and export quantities (from UN's Comtrade Database and ITC's Trade Map). We use each commodity export and its corresponding single GSCI price index to construct our weighted commodity performance measure. Exceptions are for the sub-groups wheat, cattle and energy, which are in bold type, and for which we use the GSCI sub-group price index.

3.2.3 Set of Control Variables

In order to isolate the impact of country-specific commodity performance on sovereign risk, we introduce a broad set of explanatory variables to capture international and national financial market developments.

To distinguish the effects of country-specific commodity price shocks from general economic fluctuations affecting a country, we control for a country's daily stock market returns. Stocks returns should partly capture the effects of commodity prices either via stock prices of commodity exporting companies or by signaling the overall stance of the economy. Controlling for stock returns, we aim to measure the impact of exported commodity prices on sovereign risk beyond these effects. Deriving daily, liquid stock market data for emerging markets can be challenging and therefore restricts our sample as described above. We draw equity returns from either MSCI, Datastream or S&P, depending on which provider has the longest and most complete series. We handle liquidity concerns the same way as we did for EMBIs by setting zero returns to missing if they occur for more than two consecutive trading days. All of our series are in U.S. dollar in order to match with EMBI and GSCI returns. Introducing EMBI and contemporaneous stock returns could lead to reverse causality concerns. We therefore lag the stock returns by one day, though our results do not depend on this choice. We expect higher lagged stock returns to have a negative effect on sovereign risk.

As a second country-specific control variable, we introduce exchange rate returns of each country's currency towards the U.S. dollar. Higher commodity prices could lead to an appreciation of a country's currency which could affect the export performance of non-commodity exporting firms and therefore impact sovereign risk. Exchange rate movements are measured as a daily percentage change and drawn from Thomson Reuters. We again lag the returns by one day as higher EMBI spreads could otherwise drive the exchange rate of the same day. Higher values indicate depreciation of the domestic currency against the U.S. dollar.

We further control for daily changes in the VIX to capture the implied volatility of U.S. equity markets. Also, we include the U.S. corporate credit spread which is the yield difference between the S&P U.S. high yield corporate bond index and the corresponding S&P investment grade corporate bond index. Both variables capture volatility and risk premiums in U.S. financial markets that could easily spill-over to emerging market financing conditions, given the importance of global factors for sovereign creditworthiness (Longstaff et al. 2011). We expect them to enter with a positive sign in describing EMBI spreads of a country. We also control for the U.S. term spread, i.e. the yield difference between a 10-year U.S. treasury bond and a 3-month U.S. T-Bill. The term spread approximates the premium investors receive for long-term investments. We furthermore control for changes in the yield of 10-year U.S. treasury bonds to approximate the general interest rate environment. Lastly, we want to control for general effects in the market for government debt. We do so by including the daily return of the BofA Merrill Lynch global government bond index and expect a negative correlation with emerging market sovereign risk.

All control variables are winsorized at the 1st and 99th percentile to alleviate the impact of outliers. Summary statistics for all daily variables in our baseline estimation can be found in Table 3.3, all precise variable definitions and sources can be found in Table A.3.3.

3.3 Empirical Strategy

3.3.1 Baseline Specification and Results

We estimate the following OLS panel regression model using daily data for 34 countries from $t = $ January 1, 1994 to December 31, 2016:

$$\Delta EMBI_{it} = \beta_1 \Delta CommodityPerformance_{it} + \beta_x \Delta Controls_{(i)t} + \alpha_i + \delta_{m_t} + \epsilon_{it} \quad (9)$$

$\Delta EMBI_{it}$ measures daily changes in the EMBI Global Spread of country i relative to U.S. Treasuries. Higher EMBI spreads indicate rising sovereign risk. $\Delta CommodityPerformance_{it}$ is the right-hand-side variable of interest and captures export-weighted commodity price returns of each country, as described in Section 3.2.2. Our hypothesis is that β_1 is negative, i.e. higher prices of a country's key commodity exports are associated with lower sovereign risk.

$\Delta Controls_{(i)t}$ encompasses all control variables introduced in the previous section, i.e. lagged stock returns, lagged exchange rate returns, global government bond index returns and changes in the VIX, U.S. corporate spread, U.S. term spread and 10-year U.S. treasury yields. α_i are country fixed effects which account for time-invariant country-specific factors, such as permanent market structures in a country's commodity exports. We also include time fixed effects δ_{m_t} for every month to capture time-specific, market-wide developments

Table 3.3: Summary statistics of all variables

Variable	Obs.	Mean	Median	Std. Dev.	Min	Max
ΔEMBI	148,973	-0.0808	-0.0808	9.433	-21	21.00
ΔCommodityPerformance	175,583	0.0130	0.0130	1.333	-5.226	5.108
ΔStockIndex	165,407	0.00902	0.00902	1.623	-5.478	5.187
ΔExchangeRate	196,316	0.0203	0.0203	0.543	-1.906	2.197
ΔVIX	204,000	-0.00248	-0.01	1.315	-3.875	4.755
ΔGlobalGovernmentBondIndex	204,000	0.0177	0.0126	0.377	-1.004	1.048
ΔTermSpread	204,000	-0.000149	0	0.0591	-0.161	0.186
ΔCorporateSpread	204,000	-2.30e-05	-0.00136	0.0811	-0.246	0.285
Δ10YearTreasuryYield	204,000	-0.000595	0	0.0560	-0.150	0.164
CommodityExportShare	177,928	22.91	14.27	23.47	0.319	99.65
CommodityStandardDeviation	180,313	1.225	1.130	0.505	0.0692	3.780
CommodityHHI	180,551	0.499	0.440	0.254	0.124	1.000
SpecializationAgriculture	177,928	0.170	0	0.376	0	1
SpecializationLiveStock	177,928	0	0	0	0	0
SpecializationEnergy	177,928	0.323	0	0.467	0	1
SpecializationIndustryMetals	177,928	0.0704	0	0.256	0	1
SpecializationPreciousMetals	177,928	0.0601	0	0.238	0	1
GDPGrowth	204,034	0.973	1.081	2.297	-10.83	9.019
TaxRevenueToGDP	163,848	15.74	15.38	4.643	1.830	26.49
CorporateProfitToGDP	90,145	0.252	0.208	0.229	5.91e-05	1.951
DebtToGDP	172,721	47.00	42.55	28.01	5.350	143.5
Inflation	203,740	10.19	6.180	12.02	0.570	50.04
SovereignDebtCrisis	204,000	0.0141	0	0.118	0	1
YearsSinceLastRestructuring	204,000	22.47	19	15.17	0	40
GDPPerCapita	202,699	5,600	4,765	3,808	505.1	15,202
FederalFundsRate	204,034	2.694	2	2.345	0.0400	7.800
ControlOfCorruption	186,320	42.54	45.16	19.78	0.510	91.88
RuleOfLaw	186,320	42.41	43.19	19.14	0.470	89.47
PoliticalStability	186,320	35.13	34.12	20.99	0.470	91.49
GiniRedistribution	198,815	5.381	4.200	5.985	-6.200	24.10
ManufacturingShare	196,467	16.47	15.75	5.340	6.553	31.55
FDI-Inflows	202,990	3.439	2.680	2.828	-0.250	13.90
GFCF	202,728	22.46	21.66	6.102	9.790	43.59
ReservesToGDP	199,385	0.152	0.129	0.0970	0.00110	0.453
KOF	204,034	62.14	61.97	9.699	32.06	86.10
ChinnIto	195,948	-0.0883	-0.136	1.308	-1.910	2.360
IBRDLoans	196,466	0.0381	0.0206	0.0561	0	0.394
NetAidGNI	185,512	1.080	0.350	2.062	-0.220	12.11

Sample period is 01/01/1994–31/12/2016. Variables with Δ are in daily growth rates, all other variables are in levels. ΔEMBI is winsorized at 5th and 95th, all other variables at 1st and 99th percentile. See Table A.3.3 for definitions and sources.

that have a common effect on all countries. We cluster standard errors at the country level to allow for the correlation of unobserved factors in the error terms within countries.

The results if and by how much export-weighted commodity prices affect sovereign risk are reported in Table 3.4. Model (1) uses neither control variables, nor fixed effects. Model (2) includes fixed effects. Baseline model (3) uses the full set of control variables and fixed effects. Model (4) interacts the commodity price index with a dummy variable indicating as to whether the country is a heavy commodity exporter, which is defined as having a share of commodity exports on total exports equal or above the 90th percentile. This percentile starts at a commodity export share of 56.5% and applies to Ecuador, Ghana, Kazakhstan, Nigeria, Peru and Venezuela. In all specifications, we can reject the null hypothesis of a zero effect of commodity price returns on sovereign bond returns at the 1% level of statistical significance. Investors appear to anticipate an increase in sovereign risk when the prices of a country's exported commodities deteriorate. This result has a strong footing in the literature, for example, Hilscher & Nosbusch (2010) show that commodity prices are a key determinant of a country's terms of trade, which are known to affect sovereign risk.

Turning to the economic significance, an increase in the commodity performance variable by one standard deviation is associated with a 33.2 bps (0.2488×1.3328) increase in EMBI spread differences on average which corresponds to roughly 3.5% of the standard deviation of EMBI differences ($0.3316 \div 9.4332$) which is economically meaningful (column (3)). For heavy commodity exporters (with a commodity share on total exports equal or above the 90th percentile, see column (4)) we find that a one standard deviation increase in export commodity price returns is associated with a 47.5 bps increase in EMBI spread differences (corresponding to 5% of the EMBI's standard deviation). This amount equals around 40% of the standardized effects of the VIX and the corporate bond spread. Commodity price changes are therefore an important driver of the sovereign debt performance of emerging markets in the daily perspective of financial markets.

Regarding the remaining control variables, we find signs and significance levels broadly in line with our expectations. Stock market returns enter with a negative sign in the regression and are both statistically and economically highly significant. Positive changes in the VIX, the corporate spread and exchange rate depreciations are associated with higher EMBI spreads, whereas rising global government bond returns and U.S. treasury bond yields enter with a negative sign and all with statistically significant coefficients. The term spread carries a positive sign but is not statistically significant. Our evidence for the importance of these global variables is well-embedded in the literature (Longstaff et al. 2011).

Table 3.4: Baseline results: commodity performance and sovereign risk

	(1) ΔEMBI	(2) ΔEMBI	(3) ΔEMBI	(4) ΔEMBI
ΔCommodityPerformance	-0.855***	-0.812***	-0.249***	-0.202***
	(0.0688)	(0.0669)	(0.0439)	(0.0434)
HighComExport				0.322**
				(0.147)
ΔCommodityPerformance $\times$ HighComExport				-0.357**
				(0.158)
ΔStockIndex			-0.221***	-0.223***
			(0.0348)	(0.0351)
ΔExchangeRate			0.545***	0.551***
			(0.0863)	(0.0868)
ΔVIX			0.891***	0.882***
			(0.139)	(0.138)
ΔGlobalGovernmentBondIndex			-0.349**	-0.363**
			(0.137)	(0.137)
ΔTermSpread			1.084	0.967
			(0.808)	(0.849)
ΔCorporateSpread			15.26***	15.10***
			(1.169)	(1.161)
Δ10YearTreasuryYield			-46.32***	-46.39***
			(3.465)	(3.428)
Constant	-0.0692*	-0.0697***	-0.101***	-0.135***
	(0.0381)	(0.000844)	(0.00356)	(0.0155)
Observations	143,075	143,075	129,316	128,031
R-squared	0.016	0.040	0.217	0.219
Time & Country FE	No	Yes	Yes	Yes
Number of Countries	34	34	34	34

This table shows results from OLS-panel regressions of the daily first difference of a country's Emerging Market Bond Index Spread (ΔEMBI) on the daily returns on the weighted price index of a country's exported commodities (ΔCommodityPerformance) and controls. Estimation period is from 01/01/1994 to 12/31/2016. Variable definitions are provided in Table A.3.3. HighComExport is a dummy variable being 1 if a country's share of commodity exports on total exports is equal or above the 90th percentile of countries in the panel and 0 otherwise. Estimations include country and time fixed effects on the monthly level. Standard errors (in parentheses) are clustered at the country level, ***, ** and * indicate statistical significance at the 1%, 5% and 10% level, respectively.

3.3.2 Alternative Specifications

In this section, we address possible concerns in our empirical specification. First, we consider the effects of market power of domestic commodity exporters. Most papers argue that commodity prices traded at highly centralized world markets are exogenous to domestic fundamentals (e.g. Chen & Rogoff (2003)). Still, the largest exporters of a commodity may not have to take global commodity prices as given, but can strategically manipulate raw material rates through domestic production decisions (Clements & Fry (2008)). If so, domestic concerns such as deteriorating sovereign creditworthiness could impact commodity production which would then affect global commodity prices and thus entail a reversal effect in our econometric inference. To test the importance of this concern, we specify a version of our commodity portfolio variable that is more precise in affecting only price takers of a raw material. To do so, we construct the shares of each country's commodity exports on the global export volume of this specific commodity. We then remove a commodity in the weighted portfolio of a country if this country has at any point in our sample period a global export share of more than 10% for the respective commodity. This threshold is fairly low and Table A.3.1 shows that almost every commodity is affected by one or more of such dominant global exporters. If a commodity is removed from a country's portfolio due to this procedure, the remaining commodity weights are re-adjusted. We repeat our baseline estimation with this world market-adjusted commodity performance version. The estimated coefficient of the adjusted commodity variable remains statistically highly significant and is even somewhat larger than the baseline version (Table 3.5, column (1)). This result could indeed suggest that there is strategic behavior in price-setting decisions. Nevertheless, this specification underlines our main result that emerging markets' sovereign creditworthiness is commodity-dependent, and, if anything, price-taking commodity exporters are even stronger affected.

Second, we want to make sure that the variation in our export-weighted commodity variable is not driven by re-exported commodities. Should raw materials actually be imported from other countries and then get re-exported, we would falsely classify countries as commodity exporters even though actual net exports are much lower. A related issue could be that price increases of key import goods could dominate favorable price fluctuations of important export products. To address these concerns, we construct a portfolio variable capturing the net export values of commodity sales. To this end, we first multiply a country's absolute export value (in U.S. dollars) of each commodity with its daily price change. This measure gives an indication of the extra export revenue generated or lost due to the commodity's price change. We do the same procedure for absolute import values (in U.S. dollars) and aggregate revenue changes for imports and exports for each country on a daily basis. Second, we subtract the import-weighted price changes from the export-weighted price changes. The resulting variable gives us the net-export values we are after by allowing for negative net-exports and hence negative returns, for instance if a country that imports more energy commodities than it exports faces rising energy prices. Lastly, we scale the derived net returns by dividing the net-export return variable by each country's GDP in U.S. dollars. In sum, this variable adjusts the original commodity performance by taking the price fluctuations of a country's most important import commodities into account. In this way, the variable approximates the commodity-specific terms of trade of a country, which were shown to matter for sovereign risk in Hilscher & Nosbusch (2010). Results in column (2) of Table 3.5 illustrate that the derived variable has a negative effect on sovereign risk that is statistically significant at the 1% level. Commodity prices are therefore, even when only regarding net exports, affecting the sovereign debt performance of emerging markets.

Lastly, our baseline models (plausibly) assume a continuous price impact so that a one unit change in commodity prices results in a given impact on yield spreads (no matter how large the price change in commodities). In the following robustness check, we assume that investors' attention is rather focused on trading days with large price changes in commodity prices rather than a continuous pricing. To do so, we change our commodity performance variable in a way that takes both the economic importance of exported commodities (affectedness) and key price events (treatment) into account. To this end, we mark the (at most) five commodities in a country's portfolio that have the greatest weight as long as this weight share is over 10% of total commodity exports. No country has a higher number of commodities than five for which this criterion applies. These commodities are coded with 1, other commodities with 0. We then mark all trading days in which a commodity had a positive price shock which is defined as having a price change above the respective 75th percentile (positive shock). We do the same for negative shocks, defined as a price change below the 25th percentile. We multiply the dummies for a country's most important commodities with their respective positive and negative price shock variables, separately. The resulting country-specific and daily-varying variable for each commodity is 1 if the commodity is economically important for the respective country and has a positive price shock event on this day. We then aggregate these dummies over all commodities, separately for the positive and negative price shocks. The resulting positive-shock variable can take values from 0 (no price shock for economically important commodities) to 5 (all economically important commodities for a country are subject to a positive price shock on the same day). Finally, we subtract the aggregated negative shocks from the aggregated positive shocks and arrive at a net-shock-indicator that ranges between -5 (all important negative shocks materialize) and +5 (all important positive shocks materialize). Note that a value of e.g. 0 on a given day can imply that either no price shock that mattered for the respective country took place or that occurring positive and negative shocks canceled each other out. We use the net-shock-indicator as our new commodity performance measure in our baseline. Column (3) reports a statistically significant effect at the 1% level. Economically, if the indicator increases by 1 unit, the EMBI spread differences decrease by roughly 18.7 bps.

Table 3.5: Alternative specification results

	(1) ΔEMBI	(2) ΔEMBI	(3) ΔEMBI
ΔCommodityPerformance: Excluding world-market-relevant Exporters	-0.259*** (0.0423)		
ΔCommodityPerformance: Adjusting for Imports		-1.872*** (0.556)	
ΔCommodityPerformance: NetShockIndicator			-0.187*** (0.0335)
Observations	129,901	126,483	129,901
R-squared	0.216	0.219	0.216
Number of Countries	34	34	34
Time & Country FE	Yes	Yes	Yes
Controls	Yes	Yes	Yes

This table shows results from OLS-panel regressions of the daily first difference of a country's Emerging Market Bond Index Spread (ΔEMBI) on the daily returns on the weighted price index of a country's exported commodities (ΔCommodityPerformance) and controls. Column (1) excludes a commodity from a country's portfolio if the country had at any point in time a world market share of more than 10% for this commodity (see Table A.3.1 for affected commodities). Column (2) takes imported commodities in the calculation of ΔCommodityPerformance into account. Column (3) captures relevant commodities for each country and daily price events and aggregates them to a net-shock index. Estimation period is from 01/01/1994 to 12/31/2016. Variable definitions are provided in Table A.3.3. Control variables include a country's stock index and exchange (to U.S. Dollar) returns, changes in the VIX, U.S. term spread, U.S. corporate spread, U.S. 10-year treasury yield and global government bond index. Estimations include country and time fixed effects on the monthly level. Standard errors (in parentheses) are clustered at the country level, ***, ** and * indicate statistical significance at the 1%, 5% and 10% level, respectively.

3.4 Drivers of the Commodity-Sovereign Risk Dependence

We now turn to investigate potential drivers of the spillover of export-weighted commodity price changes to sovereign risk. If emerging markets are commodity-dependent, as the previous section indicated, it is important for policy makers to know what affects this dependency and which macroeconomic factors or policy measures can potentially reduce commodity dependence. These policy-parameters are so far hardly explored. For instance, Bouri et al. (2017) find that the spillover of commodity prices on sovereign risk varies across countries and over time, which the authors argue to be due to politically, economically or monetary policy related factors. We intend to identify these driving factors. We test the hypotheses if commodity-sovereign risk is driven by channels approximating commodity-related factors (3.4.1), macroeconomic and international factors (3.4.2), and a range of possible policy measures to limit commodity dependence (3.4.3). For each channel under investigation we estimate the baseline regression (9) and interact, in order to rule out reverse effects, with the yearly-, quarterly-, monthly- or daily-lagged value of the respective channel unless stated otherwise:[26,27]

$$\Delta EMBI_{it} = \lambda_1 \Delta CommodityPerformance_{it} \times Channel_{it-1} + \lambda_2 Channel_{it-1} +$$
$$\lambda_3 \Delta CommodityPerformance_{it} + \lambda_x \Delta Controls_{(i)t} + \alpha_i + \delta_{m_t} + \epsilon_{it} \tag{10}$$

We expect channels that increase the commodity dependence of emerging markets to enter with a negative sign for the respective interaction term, while channels that could mitigate the spillover to have a positive interaction coefficient.

Following Nizalova & Murtazashvili (2016) and Bun & Harrison (2019), we argue that our interaction coefficients are consistently estimated, as long as one variable in the interaction term is exogenously determined. This assumption holds plausibly for weighted commodity prices which are largely world-marked determined. Furthermore, we demonstrated in Section 3.3.2 that potentially biasing effects are small in size and do not disturb our main results. Therefore, even if some channels could be endogenous with respect to sovereign risk, we argue that the interaction terms allow for exogenous interpretation. Summary statistics on interaction variables are in Table 3.3, all data sources are reported in Table A.3.3.

3.4.1 Commodity-related Factors

A natural starting point is to check if countries that have a larger share of commodity exports on their total export volume also face a more forceful commodity price spillover. To test this, we interact with the share of total commodity exports on total exports of each coun-

---

[26]Since we use an emerging market panel, not all countries have full data on all interaction variables. We report on this when it becomes an issue. Definition and sources of all variables can be found in Table A.3.3

[27]In order to ensure correct specification, interaction models also contain the single linear terms of interacted variables (Brambor et al. 2006).

try. As expected, the interaction coefficient is negative and statistically significant at the 5% level as reported in Table 3.6, column (1). The margin plot depicted in Figure 3.1 shows the marginal effect of increasing commodity performance on sovereign risk depending on the level of the commodity-export share. It suggests that commodity price changes turn statistically significant in impacting sovereign risk at an export share of roughly 5% and increase their impact further beyond this threshold.[28] Thus, a larger commodity export industry increases the commodity-sovereign risk dependence. Possible explanations for this result may be that a larger commodity export industry is associated with a more pronounced impact of commodity prices on economic growth and thus fiscal revenues and public expenditures. A larger commodity export industry should also increase potential direct cash flows to the government via publicly owned commodity firms or royalties to the government.

We next test if higher volatility of a country's export commodity prices is associated with a more intense commodity-sovereign risk dependence. To do so, we calculate the rolling standard deviation of each country's export-weighted commodity returns on a 23-day basis, which is roughly the number of trading days each month. The respective interaction coefficient is, however, statistically insignificant (column (2)). In addition, the resulting margin plot reported in Figure 3.1, does not lead to the conclusion that with more volatile price fluctuations of a country's key commodities in the previous month, current price changes have stronger effects on sovereign risk.

Having a high concentration in just one commodity could be associated with a stronger commodity dependence of a country since it has no diversification benefits in case of a shock to its key raw material. We test this hypothesis by constructing the yearly Hirschman-Herfindahl-Index (HHI), i.e. the sum of squared commodity export weights for each country. The HHI varies from roughly 0.15 for well-diversified export countries such as Poland, to almost 1 for oil-exporting countries such as Nigeria or Venezuela. The interaction of the contemporaneous HHI with the commodity portfolio yields a coefficient with positive sign that is, however, small and statistically insignificant (Table 3.6, column (3)). This ambiguous relationship is also confirmed in the margin plot depicted in Figure 3.1 and was also found similarly in the latest report by UNCTAD (2019). One explanation of this result could be that

---

[28]The GSCI, from which we derive the included commodities, covers the most important but not all commodities. It is unlikely that our results are biased because of this, nevertheless, calculated ratios such as commodity export shares are not comparable one-to-one with those reported e.g. in UNCTAD (2019).

as long as a country is commodity-dependent, it does not matter much if this dependency is towards several or only one raw material. We therefore conclude that the volatility, variety or concentration of commodities is only of secondary importance for understanding commodity-sovereign risk dependence of emerging markets.

Next, we test if exporting certain commodities entails stronger commodity dependence than others. We focus on the commodity subgroup level depicted in Table 3.2. We define a country as being specialized in a certain commodity subgroup if this subgroup has a share on total exports above the 10% ratio. This criterion has the advantage that more diversified countries can be specialized in several commodity subgroups. Also, we capture economically critical specialization patterns. Interacting with the five dummies representing the subgroups, we find that commodity price dependence is slightly larger for countries specializing in energy with the interaction term statistically significant at the 10% level (column (4)). Bouri et al. (2017) and Bouri et al. (2018) find a similar effect. On the other hand, countries specializing in exporting industrial metals reveal statistically significantly lower commodity dependence at the 5% level. One reason could be that industrial metal sectors could facilitate the fostering of manufacturing sectors and other downstream technologies, which we show to be important drivers in reducing commodity dependence in the following sections. Another explanation could be that the public sector share of energy exports (such as crude oil) is larger compared to the public sector share in exporting industrial metals.

Table 3.6: Drivers of commodity-sovereign risk dependence: commodity-related factors

	(1) ΔEMBI	(2) ΔEMBI	(3) ΔEMBI	(4) ΔEMBI
ΔCommodityPerformance	-0.0824	-0.290***	-0.351**	-0.151***
	(0.0656)	(0.0751)	(0.142)	(0.0548)
CommodityExportShare	0.0157*			
	(0.00784)			
ΔCommodityPerformance	-0.00599**			
$\times$ CommodityExportShare	(0.00284)			
CommodityStandardDeviation		-0.318***		
		(0.0833)		
ΔCommodityPerformance		0.0217		
$\times$ CommodityStandardDeviation		(0.0485)		
CommodityHHI			-0.170	
			(0.249)	
ΔCommodityPerformance			0.158	
$\times$ CommodityHHI			(0.253)	
SpecializationAgriculture				-0.00725
				(0.0902)
ΔCommodityPerformance				-0.190
$\times$ SpecializationAgriculture				(0.161)
SpecializationEnergy				0.324
				(0.239)
ΔCommodityPerformance				-0.193*
$\times$ SpecializationEnergy				(0.0981)
SpecializationIndustryMetals				-0.0364
				(0.154)
ΔCommodityPerformance				0.393**
$\times$ SpecializationIndustryMetals				(0.181)
SpecializationPreciousMetals				0.523**
				(0.194)
ΔCommodityPerformance				-0.510
$\times$ SpecializationPreciousMetals				(0.331)
Observations	128,478	128,097	128,031	128,031
R-squared	0.220	0.218	0.218	0.219
Number of Countries	34	34	34	34
Time & Country FE	Yes	Yes	Yes	Yes
Controls	Yes	Yes	Yes	Yes

This table shows results from OLS-panel regressions of the daily first difference of a country's Emerging Market Bond Index Spread (ΔEMBI) on the daily returns on the weighted price index of a country's exported commodities (ΔCommodityPerformance) and controls. Interaction terms of ΔCommodityPerformance with CommodityExportShare ((1), share of commodity exports on total exports), CommodityStandardDeviation ((2), rolling standard deviation of ΔCommodityPerformance of past 23 business days), CommodityHHI ((3), concentration index of export weights in ΔCommodityPerformance), specialization in different commodity subgroups ((4) subgroup exports are at least 10% of total exports) are estimated. Estimation period is from 01/01/1994 to 12/31/2016. Variable definitions are provided in Table A.3.3. Control variables include a country's stock index and exchange rate (to U.S. Dollar) returns, changes in the VIX, U.S. term spread, U.S. corporate spread, 10-year U.S. treasury yield and global government bond index. Estimations include country and time fixed effects on the monthly level. Standard errors (in parentheses) are clustered at the country level, ***, ** and * indicate statistical significance at the 1%, 5% and 10% level, respectively.

Figure 3.1: Marginal effects of commodity performance on EMBI spread returns interacted with commodity-related factors. Bars indicate 95% confidence intervals. Distribution of interaction variable is shown. Results of the corresponding regressions are in Table 3.6.

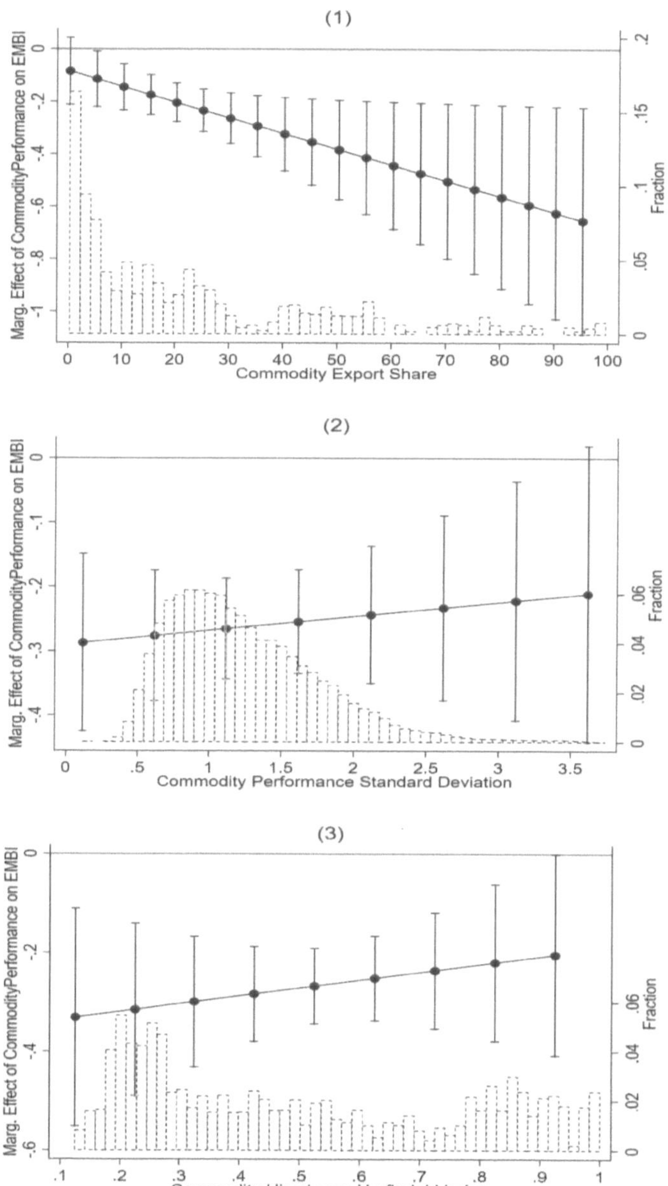

3.4.2 Macroeconomic and International Factors

We turn to investigate the impact of broader macroeconomic factors with respect to the commodity dependence of emerging markets. Different macroeconomic environments could ease or strengthen a country's dependence on its commodities. For instance, one could hypothesize that during a business cycle downturn, income streams from raw materials matter more for countries than in good times because they can provide the fiscal means to address the recession. We therefore start by interacting the export-weighted commodity portfolio with lagged GDP growth, measured on a quarter-to-quarter basis. The resulting interaction term, reported in column (1) of Table 3.7, enters with a positive sign that is statistically significant at the 10% level. The margin plot in Figure 3.2 additionally confirms that prices of exported commodities matter more for countries if they are in a business cycle downturn. Commodity price spillovers turn insignificant at a quarterly GDP growth level of roughly 4%. This result fits into the general finding in the literature, that asset price co-movement is intensified during crisis periods (see, e.g., Hartmann et al. (2004)).

We dig deeper into the importance of the business cycle, first, by interacting with the tax revenues of a government scaled by GDP. Since tax revenues vary positively with the business cycle but also, if higher, make a country less dependent on export gains from commodities, we would also expect a positive coefficient for the interaction term of lagged tax revenues and weighted commodity price changes. We find some confirmation for this with the positive margin plot in Figure 3.2 in which commodity shocks become statistically insignificant in explaining sovereign risk at higher levels of tax revenues. The corresponding coefficient in column (2) is positive but not statistically significant. One further measure for business cycle effects are profits achieved in the corporate sector. We therefore interact our commodity variable with the lagged ratio of corporate sector profits to GDP which is, however, only available for 16 countries in our sample. The resulting coefficient has the expected positive sign and is statistically significant at the 10% level (column (3)). In addition, the margin plot depicted in Figure 3.2 lends support to the hypothesis that with higher corporate profits, commodity price spillovers eventually matter less than in times of lower private profits. Taking the reduced sample size into account, we interpret these first three estimations as evidence that commodity price shocks hit countries harder if their business cycle is in downturn and if both private and public sectors have less capacity in terms of profits or tax revenues to fend off negative shocks. Our result is connected to Aizenman et al. (2013) who find a country's fiscal space to be important for its sovereign risk level.

The indebtedness of a country could be important for its reliance on commodities. Export gains from raw materials might matter more for a country as an income source to stabilize debt ratios if sovereign debt is larger which would speak for increased commodity dependence. When interacting with the lagged debt-to-GDP ratio of a country, we find, however, a positive interaction coefficient that is weakly statistically significant (column (4)). A possible explanation for this surprising result may be that countries with on average higher public debt ratios are less reliant on commodity exports because they can issue even more debt to buffer commodity price shocks due to their higher fiscal capacities. On the contrary, countries with low debt ratios may already have reached their country-specific critical debt levels and may therefore not be able to issue new debt.

We next test the hypothesis that higher rates of inflation could be linked to commodity price spillovers of emerging markets. If money loses its purchasing power through inflation, income gains from commodities that are measured in U.S. dollar might matter more to stabilize sovereign creditworthiness. However, we find no empirical confirmation for this hypothesis. We report an estimated interaction term with lagged annual inflation (column (5)) that is negative but statistically insignificant. The margin plot depicted in Figure 3.2 supports this finding. Though other papers like Aizenman et al. (2016) find inflation to be an important determinant of sovereign risk, it could be the case that inflation dynamics do not work through commodity prices in achieving this impact.

Related to the previous interactions, we test if commodity dependence increases if a country suffers a sovereign debt crisis. To this end, we exploit the systemic banking crises database by Laeven & Valencia (2018). We interact commodity performance with a contemporaneous dummy that indicates the year in which a country had a sovereign debt crisis. However, with results shown in column (6), we find only weak confirmation that commodity price shocks have a stronger spillover on sovereign creditworthiness during a sovereign debt crisis. While we find the expected negative interaction coefficient, it is statistically insignificant. One reason for this could be measurement error in that the crisis dummy is on a yearly basis which is too imprecise given the daily frequency of our data.

Digging deeper into the sovereign repayment history of a country and using an approach that is less susceptible to the data issue above, we interact with a continuous variable that measures the number of years since the last debt restructuring event occurred. We also include those restructuring events that happened before the start of our sample period in 1994. Overall, 22 countries in our panel negotiated at least one sovereign debt restructuring. The highest number of years since the last restructuring event is 36. For the twelve non-defaulters, we therefore set the variable to 40 as a measure for a sovereign repayment history without any restructuring events. The continuous variable enters negatively and statistically significantly at the 10% level in interaction with commodity price changes (column (7)). The margin plot in Figure 3.2 furthermore confirms the hypothesis that a country with a distant or no sovereign debt restructuring history is hit significantly less by price shocks of its commodity exports compared to a country with only recent cases of bond renegotiations. This result could imply that financial markets pay closer attention to the commodity price performance of countries with a less stable debt repayment history in recent years as suggested by Reinhart et al. (2003), so that, for instance, negative price shocks of key commodities also have a more forceful impact on the riskiness of the respective country's debt.

Next, we test if the level of economic development matters for commodity related spillovers. To this end, we build an interaction term between commodity performance and GDP per capita of each country. The resulting interaction term has a negative sign but is small and statistically insignificant (column (8)). This result could indicate that with regard to the within variation of economic development that we are capturing, commodity dependence is sticky for emerging market economies even if a country grows in terms of GDP. It could also be because the countries in our sample are somewhat more developed since they report EMBI and stock market data which leaves out poorer countries e.g. in Sub-Sahara-Africa. Nevertheless, this result gives us some confirmation that our remaining results are not driven by any biases between richer and poorer countries, e.g. when it comes to institutional characteristics that could be a function of economic development.

	(1) ΔEMBI	(2) ΔEMBI	(3) ΔEMBI	(4) ΔEMBI	(5) ΔEMBI	(6) ΔEMBI	(7) ΔEMBI	(8) ΔEMBI	(9) ΔEMBI
ΔCommodityPerformance	-0.278***	-0.399**	-0.506***	-0.464***	-0.211**	-0.244***	-0.402***	-0.0757	-0.356***
	(0.0508)	(0.156)	(0.114)	(0.142)	(0.0825)	(0.0447)	(0.0863)	(0.115)	(0.0676)
GDP-Growth	-0.00802								
	(0.0181)								
ΔCommodityPerformance × GDP-Growth	0.0343*								
	(0.0181)								
TaxRevenueToGDP		0.0307							
		(0.0208)							
ΔCommodityPerformance × TaxRevenueToGDP		0.0108							
		(0.0100)							
CorporateProfitToGDP			-2.020*						
			(1.042)						
ΔCommodityPerformance × CorporateProfitToGDP			0.821						
			(0.482)						
DebtToGDP				0.00473					
				(0.00642)					
ΔCommodityPerformance × DebtToGDP				0.00508*					
				(0.00283)					
Inflation					0.000909				
					(0.000559)				
ΔCommodityPerformance × Inflation					-0.000565				
					(0.00655)				
SovereignDebtCrisis					-0.000565				
					(0.00655)				
SovereignDebtCrisis × ΔCommodityPerformance						-0.229			
						(0.219)			
YearsSinceLastRestructuring							0.0426***		
							(0.00797)		
ΔCommodityPerformance × YearsSinceLastRestructuring							0.00708*		
							(0.00383)		
GDP-PerCapita								3.42e-06	
								(4.31e-05)	
ΔCommodityPerformance × GDP-PerCapita								-2.65e-05	
								(2.04e-05)	
Federal Funds Rate									0.217
									(0.210)
ΔCommodityPerformance × FederalFundsRate									0.0562***
									(0.0147)
Observations	129,316	108,238	72,211	120,928	128,478	129,316	129,316	129,316	129,316
R-squared	0.217	0.226	0.208	0.229	0.219	0.217	0.217	0.217	0.217
Number of Countries	34	33	16	33	34	34	34	34	34
Time & Country FE	Yes	Yes	Yes	Yes	Yes	Yes	Yes	Yes	Yes
Controls	Yes	Yes	Yes	Yes	Yes	Yes	Yes	Yes	Yes

This table shows results from OLS-panel regressions of the daily difference of a country's Emerging Market Bond Index Spread (ΔEMBI) on the daily returns on the weighted price index of a country's exported commodities·(ΔCommodityPerformance) and controls. Interaction terms of ΔCommodityPerformance with GDP-Growth (1), TaxRevenueToGDP ((2), government tax revenue to GDP), CorporateProfitToGDP ((3), corporate sector profits to GDP), DebtToGDP ((4), gross government debt to GDP), Inflation ((5), consumer price index increase), SovereignDebtCrisis ((6), dummy for Laeven & Valencia (2018) sovereign debt crisis), YearsSinceLastRestructuring ((7), number of years since last sovereign debt restructuring, 40 if no restructuring occurred), GDP-PerCapita (8) and FederalFundsRate ((9), U.S. federal funds rate) are estimated. Estimation period is from 01/01/1994 to 12/31/2016. Variable definitions are provided in Table A.3.3. Control variables include a country's stock index and exchange rate (to U.S. Dollar) returns, changes in the VIX, U.S. term spread, U.S. corporate spread, 10-year U.S. treasury yield and global government bond index. Estimations include country and time fixed effects on the monthly level. Standard errors (in parentheses) are clustered at the country level, ***, ** and * indicate statistical significance at the 1%, 5% and 10% level, respectively.

Lastly, we want to analyze the effect of U.S. monetary policy on commodity dependence of emerging markets. Interest rates set by the Federal Reserve are determined with regard to the U.S. economy and likely only partially driven by economic developments of emerging markets or commodity prices. However, as shown by Bräuning & Ivashina (2020), monetary policy decisions in the U.S. have powerful effects for emerging markets in that expansionary measures by the Federal Reserve increase international capital flows and borrowing behavior by foreign firms. Furthermore, Frankel (2006) argues that U.S. monetary policy affects the decision for commodity exporters when to extract raw materials, to hold inventories or for investors to go into emerging markets rather than U.S. treasury bills. We therefore hold the hypothesis that more expansionary monetary policy is associated with increasing commodity dependence. Our interaction coefficient in column (9) that shows the effect of commodity performance depending on the U.S. federal funds rate is in line with this hypothesis and the literature. We find a positive and highly statistically significant coefficient at the 1% level and a margin plot in Figure 3.2 which suggests that commodity dependence increases significantly at a federal funds rate lower than 5%. While we cannot say more on the precise channel, one interpretation could be that looser global financing conditions and capital flows are used by emerging economies predominantly to increase economic activity in commodity sectors.

Figure 3.2: Marginal effects of commodity performance on EMBI spread returns interacted with macroeconomic and international factors. Bars indicate 95% confidence intervals. Distribution of interaction variable is shown. Results of the corresponding regressions are in Table 3.7.

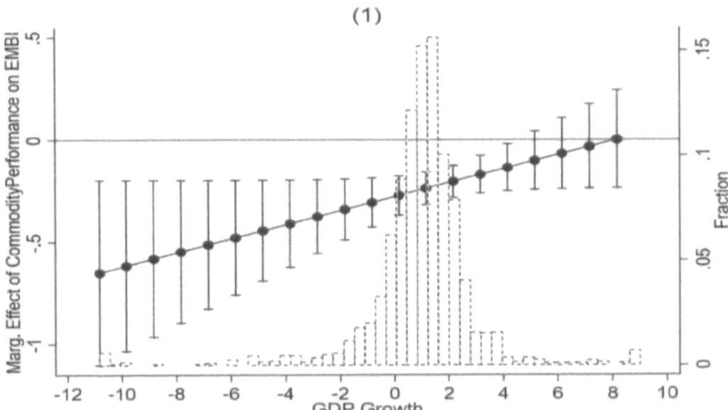

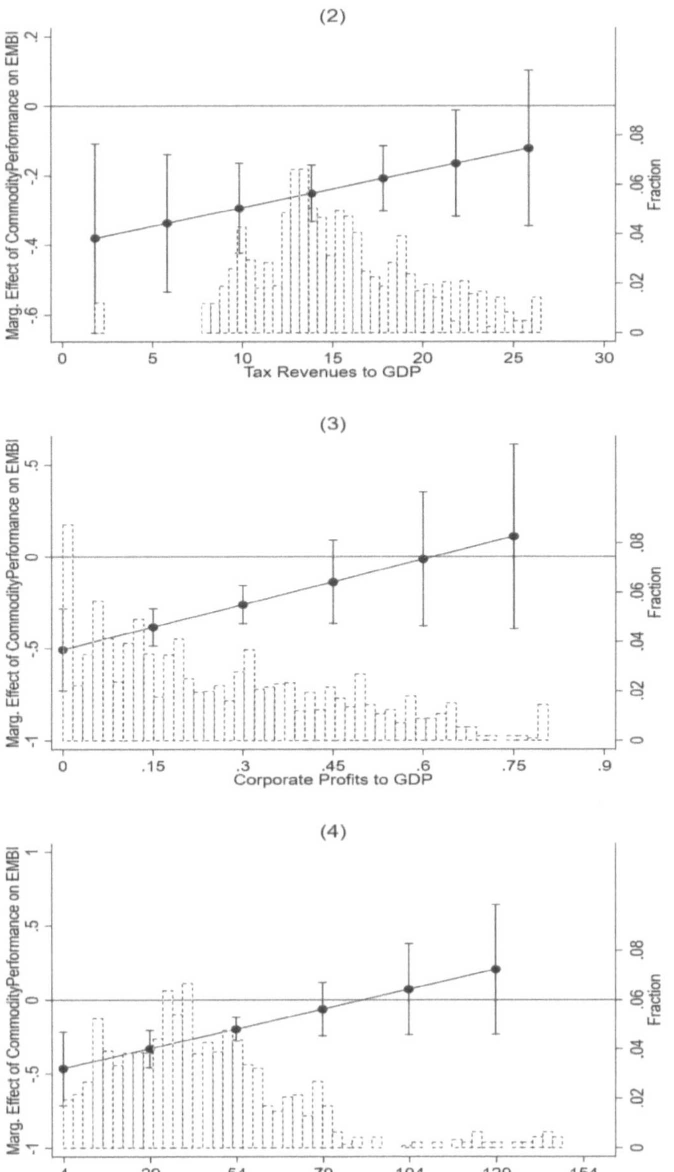

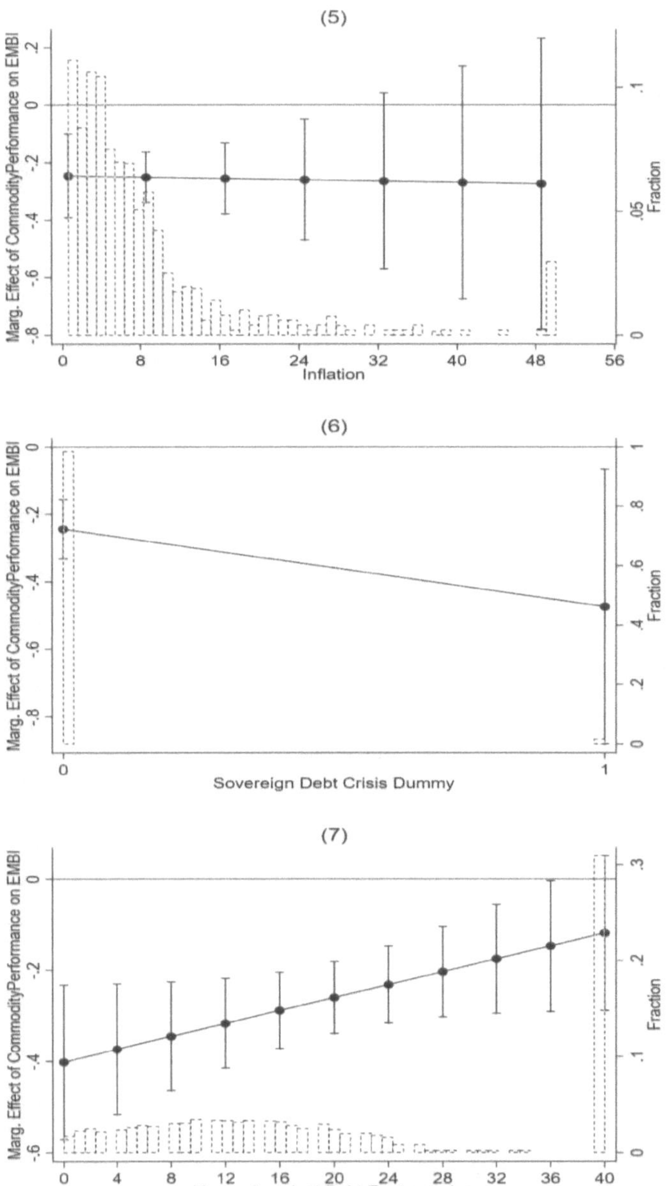
(5)
Marg. Effect of CommodityPerformance on EMBI
.2
0
-.2
-.4
-.6
-.8
Fraction
.1
.05
0
0
8
16
24
32
40
48
56
Inflation
(6)
Marg. Effect of CommodityPerformance on EMBI
0
-.2
-.4
-.6
-.8
Fraction
1
.8
.6
.4
.2
0
0
1
Sovereign Debt Crisis Dummy
(7)
Marg. Effect of CommodityPerformance on EMBI
0
-.2
-.4
-.6
Fraction
.3
.2
.1
0
0
4
8
12
16
20
24
28
32
36
40
Years since last Debt Restructuring

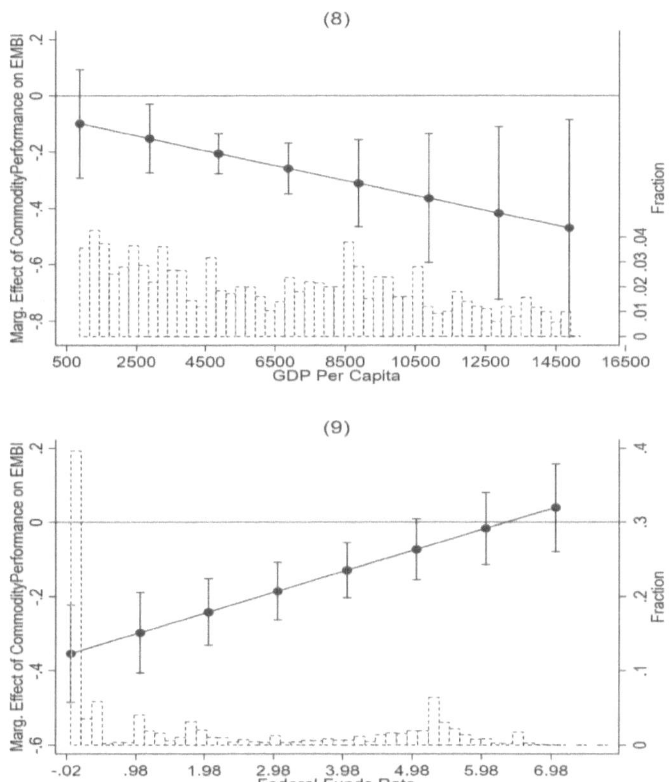

3.4.3 Policy Measures against Commodity Dependence

In order to inform the policy debate, we want to analyze what our model suggests to be promising ways to lower commodity dependence. We focus on policy measures that are to some degree more under government control than the broader macroeconomic or international variables tested above.

First, we want to investigate if countries with higher institutional quality are less commodity-dependent. To this end, we draw yearly data from the World Bank Governance Indicators which conduct extensive surveys to approximate different forms of institutional quality. We draw three indicators which we hypothesize to be related to the extraction process of commodities or the usage of incomes from commodity sales and thus to the spillover of commodity prices on sovereign risk: control of corruption, rule of law and political stability (and absence of violence), which are all available from 1996 onwards.[29]

When interacting separately with the lagged yearly values of the three measures for institutional quality, we find clear results: All interaction terms are positive and strongly statistically significant, with control of corruption and rule of law at the 1% level and political stability at the 5% level (Table 3.8, columns (1)-(3)). The margin plots depicted in Figure 3.3 support the hypothesis that with better institutional quality, commodity price shocks are less effective in impacting sovereign risk. This result implies that countries are more commodity-dependent if institutional quality is worse, for instance when ownership or legal frameworks in the production process of raw materials are less clearly structured. These findings could indicate that with improving control of corruption and a stronger rule of law, countries can mitigate rent extraction behavior in the production and selling of raw materials, reinvest gains from commodity exports more effectively, or smooth negative commodity price shocks. Our results are in line with Mehlum et al. (2006) who suggest that institutional quality is the decisive criterion for commodities to be a curse or a source of wealth.

One could be concerned that institutional quality is more difficult to improve for commodity-exporting countries (Arezki & Brückner 2011a). While our daily data structure in which institutional quality can be considered as given alleviates this concern somewhat, we also test if our results hold if we repeat the analysis for heavy commodity exporters, defined as having a commodity export share of more than 10% on total exports. We find that the statistical significance of our result remains, except for the political stability interaction (Table A.3.2). Therefore, even among stronger commodity exporters, those with effective institutions seem to fare better in terms of reducing commodity dependence.

We test the differentiating impact of commodity prices on sovereign creditworthiness on a further variable that approximates institutional quality namely the progressiveness of the tax system. We draw yearly data on the Gini coefficient of emerging markets from the database by Solt (2019).[30] We build an interaction term with commodity performance and the amount of tax redistribution, i.e. the difference between the pre- and post-tax Gini indices. This

---

[29]The indicators, ranging from 0 to 100, have occasional gaps in the early years which we close by linearly interpolating the series.

[30]We aware that data on inequality of emerging markets is imperfect, even though the data quality by Solt (2019) is considered to be standardized as best as possible. See Lang & Tavares (2018) for a discussion.

measure approximates the progressiveness of the tax system and how much elites are taxed, which could be related to the overall progressiveness of a country's institutional framework. The interaction term enters positive as shown in column (4) but is marginally insignificant before the 10% level. However, the margin plot in Figure 3.3 additionally suggests that more progressive tax systems are associated with less commodity dependence.

Next, we test three interactions that might alleviate emerging markets' commodity dependence. In a direct way, building stronger manufacturing sectors should lead to less dependence on global price fluctuations of exported commodities. In a more indirect manner, attracting FDI inflows can lead to technological spillovers which could also improve the economic structure of a country beyond pure commodity exporting. Lastly, increasing gross fixed capital formation (GFCF) i.e. investments in plant, machinery, schools and infrastructure could also diversify the economic structure of a country. We therefore interact, separately, with lagged manufacturing value-added, net FDI inflows and GFCF investments, all as a share of GDP. All interaction coefficients (columns (5), (6) (7)) are positive, with manufacturing and FDI statistically significant at the 5% level but GFCF being statistically insignificant. Still, all margin plots in Figure 3.3 strongly support the conclusion, speaking more broadly, that fostering downstream production, investing in infrastructure and technology and diversifying economic structures can be promising ways to reduce commodity dependence.

Figure 3.3: Marginal effects of commodity performance on EMBI spread returns interacted with policy measures against commodity dependence (1). Bars indicate 95% confidence intervals. Distribution of interaction variable is shown. Results of the corresponding regressions are in Table 3.8.

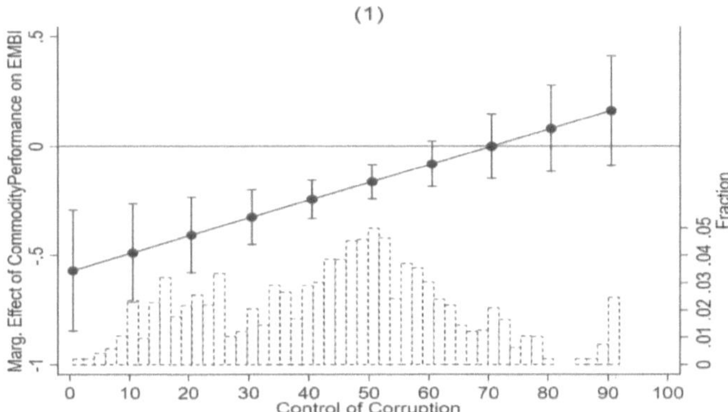

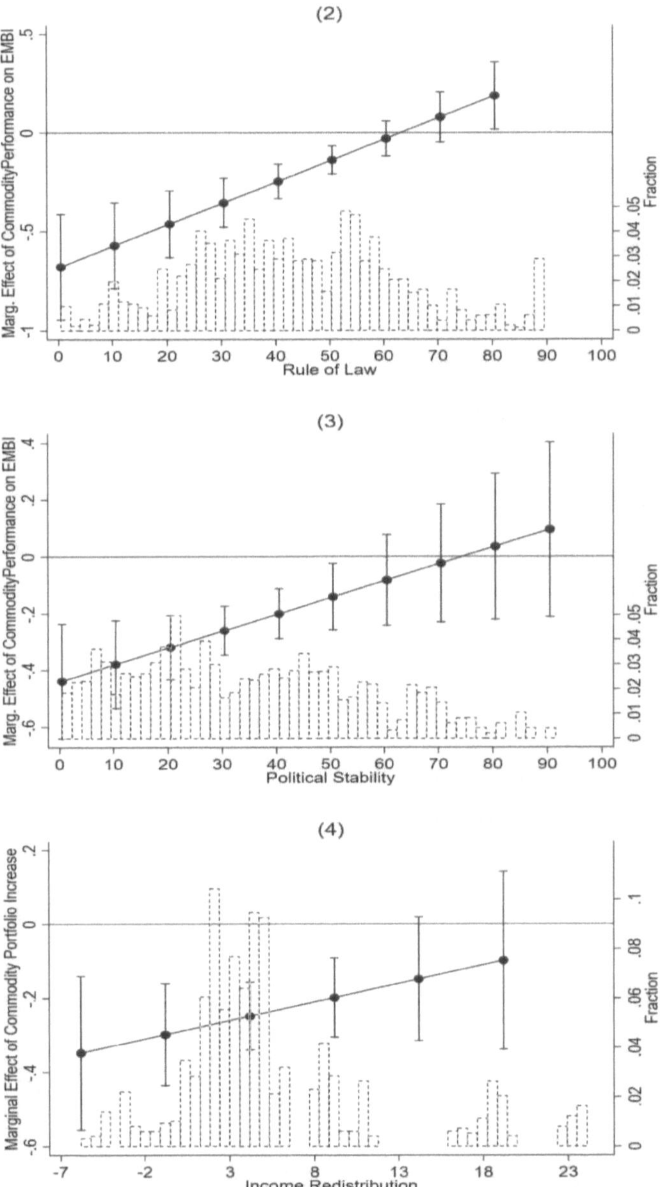

(2)
Marg. Effect of Commodity Performance on EMBI
Rule of Law
Fraction
(3)
Marg. Effect of Commodity Performance on EMBI
Political Stability
Fraction
(4)
Marginal Effect of Commodity Portfolio Increase
Income Redistribution
Fraction

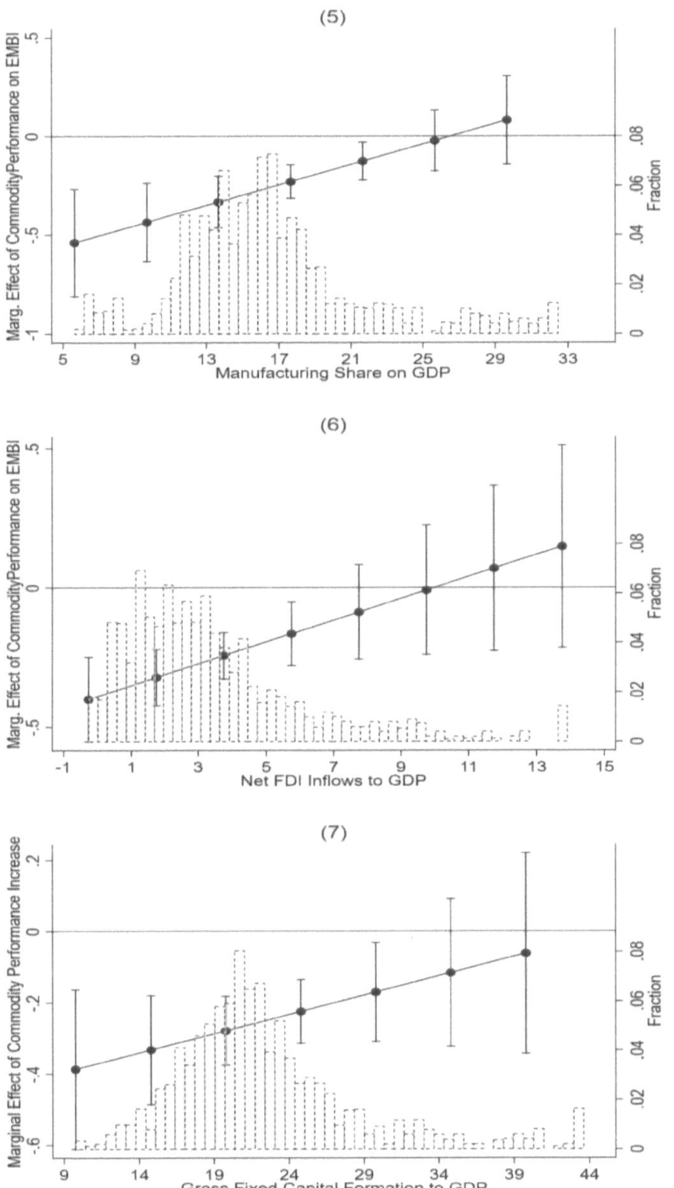
(5)
Marg. Effect of CommodityPerformance on EMBI
Fraction
Manufacturing Share on GDP
(6)
Marg. Effect of CommodityPerformance on EMBI
Fraction
Net FDI Inflows to GDP
(7)
Marginal Effect of Commodity Performance Increase
Fraction
Gross Fixed Capital Formation to GDP

Table 3.8: Drivers of commodity-sovereign risk dependence: policy measures (1)

	(1) ΔEMBI	(2) ΔEMBI	(3) ΔEMBI	(4) ΔEMBI	(5) ΔEMBI	(6) ΔEMBI	(7) ΔEMBI
ΔCommodityPerformance	-0.573***	-0.684***	-0.441***	-0.297***	-0.685***	-0.391***	-0.493**
	(0.143)	(0.137)	(0.104)	(0.0619)	(0.194)	(0.0735)	(0.190)
ControlOfCorruption	-0.000899						
	(0.00408)						
ΔCommodityPerformance × ControlOfCorruption	-0.000979***						
	(0.00285)						
RuleOfLaw		0.000388					
		(0.00346)					
ΔCommodityPerformance × RuleOfLaw		0.0108***					
		(0.00264)					
PoliticalStability			-0.000528				
			(0.00504)				
ΔCommodityPerformance × PoliticalStability			0.00594**				
			(0.00272)				
GiniRedistribution				-0.106			
				(0.113)			
ΔCommodityPerformance × GiniRedistribution				0.0109			
				(0.00738)			
ManufacturingShare					0.0197		
					(0.0180)		
ΔCommodityPerformance × ManufacturingShare					0.0258**		
					(0.00997)		
FDI-Inflows						0.00345	
						(0.0102)	
ΔCommodityPerformance × FDI-Inflows						0.0392**	
						(0.0175)	
GFCF							0.0198***
							(0.00644)
ΔCommodityPerformance × GFCF							0.0109
							(0.00811)
Observations	124,720	124,720	124,720	127,769	125,061	129,316	129,057
R-squared	0.240	0.240	0.239	0.216	0.225	0.217	0.218
Number of Countries	34	34	34	34	34	34	34
Time & Country FE	Yes	Yes	Yes	Yes	Yes	Yes	Yes
Controls	Yes	Yes	Yes	Yes	Yes	Yes	Yes

This table shows results from OLS-panel regressions of the daily difference of a country's Emerging Market Bond Index Spread (ΔEMBI) on the daily returns on the weighted price index of a country's exported commodities (ΔCommodityPerformance) and controls. Interaction terms of ΔCommodityPerformance with ControlOfCorruption ((1), World Bank control of corruption rank), RuleOfLaw ((2), World Bank rule of law rank), Political Stability ((3), World Bank political stability rank), GiniMarket ((4), difference in post- and pre-tax Gini index), ManufacturingShare ((5), share of manufacturing value-added on GDP), FDI-Inflows ((6), net FDI inflows to GDP), and GFCF ((7), gross fixed capital formation to GDP) are estimated. Estimation period is from 01/01/1994 to 12/31/2016. Variable definitions are provided in Table A.3.3. Control variables include a country's stock index and exchange rate (to U.S. Dollar) returns, changes in the VIX, U.S. term spread, U.S. corporate spread, 10-year U.S. treasury yield and global government bond index. Estimations include country and time fixed effects on the monthly level. Standard errors (in parentheses) are clustered at the country level. ***, ** and * indicate statistical significance at the 1%, 5% and 10% level, respectively.

In recent years, several countries have started to build up foreign-exchange reserves as a buffer e.g. for balance of payment crises. Higher foreign reserves could also affect commodity dependence by providing sufficient security against balance of payment crises and hence more stable macroeconomic environments. Negative terms of trade shocks, for instance through lower commodity export prices, can also be better absorbed with higher reserves. Testing this channel, we find evidence that countries with higher monthly foreign exchange reserves (relative to their GDP) are significantly less commodity-dependent as suggested by the positive and statistically significant coefficient (1%) in column (1) in Table 3.9 and the margin plot in Figure 3.4. Higher foreign exchange reserves could indeed reduce the dependence on foreign exchange inflows via exporting commodities and serve as a buffer for commodity-induced terms of trade shocks as suggested by Aizenman et al. (2012), thus mitigating the emergence of sovereign distress caused by balance of payments problems.

Furthermore, we investigate the effect of capital controls and trade openness in association with commodity price changes on sovereign creditworthiness. Aizenman et al. (2016) find trade openness to be one of the key factors in determining sovereign risk. We hypothesize that trade and capital openness is also related to commodity exports, for instance, by providing transparent market access for foreign buyers or by making sure international capital inflows can finance promising commodity projects that are difficult to fund domestically. We first use the yearly KOF globalization index by Gygli et al. (2019) as an interaction term. This index measures along several dimensions how open a country is towards trade and international financial flows. Our evidence suggests that more open countries are significantly less dependent on the price performance of their exported commodities, as shown by a significant (10% level, column (2)) and positive interaction term coefficient and the margin plot in Figure 3.4. Disentangling the KOF index into the de facto and de jure version shows that the de facto variation matters more for this effect, which enters positive and statistically significant at the 5% level (Table A.3.2). This result is in line with the findings of Aizenman et al. (2016) who underline the importance of trade openness for sovereign risk.

When using the Chinn-Ito-Index from Chinn & Ito (2006) as a measure for current and capital account openness instead of the KOF index, we find largely similar if somewhat weaker results. Though the interaction effect is positive but statistically insignificant, the marginal effect depicted in Figure 3.4 lends support to the hypothesis that more closed-off economies have a stronger dependency on their commodities for their sovereign risk, as the marginal effect of such spillovers decreases and eventually turns insignificant the more open capital accounts are. This result could suggest that more open economies could be able to better fend off a negative shock to their commodity performance because of deeper financial markets and a broader set of financing choices. The stronger effects of the de facto KOF could imply that attracting trade flows and financial investments can be a further means for diversifying economic structures away from pure commodity extraction.

Lastly, we want to investigate the effects of development assistance measures on commodity dependence. Development assistance is targeted to reduce poverty or improve health and education systems. Theoretically, more development funding could also affect commodity dependence by diversifying economic activities, incentivizing institutional reforms or investing in infrastructure projects which we showed previously as effective ways to reduce commodity dependence. We therefore interact commodity performance, first, with a country's yearly exposure of loans to the International Bank for Reconstruction and Development (IBRD) and the International Development Association (IDA) scaled to GDP. Both institutions are the main World Bank entities that extend loans to spur economic activity and to fight poverty (see Dreher et al. (2019) for a paper on the political economy of IBRD). Second, we interact with the yearly amount of net development assistance received scaled to GNI. However, we find only weak confirmation that development assistance or World Bank loans are promising ways to reduce raw material reliance of emerging markets. For both interactions, the coefficient has the expected positive sign, i.e. more assistance tends to decrease commodity dependence (columns (4) and (5)). But both coefficients are statistically insignificant and the slopes of the interaction effects, depicted in Figure 3.4, are small. If anything, we find stronger effects for IBRD loans, in that a country is more commodity-dependent if it has none or only small loan exposure compared to countries that have at least some IBRD loan exposure. Therefore, we conclude that development assistance can potentially impact commodity dependence, however, the more promising results were with regard to improving institutional quality, broadening economic structures, building up reserves and opening trade and financial accounts.

Table 3.9: Drivers of commodity-sovereign risk dependence: policy measures (2)

	(1) ΔEMBI	(2) ΔEMBI	(3) ΔEMBI	(4) ΔEMBI	(5) ΔEMBI
ΔCommodityPerformance	-0.444***	-1.059**	-0.257***	-0.335***	-0.271***
	(0.0935)	(0.434)	(0.0458)	(0.0989)	(0.0664)
ReservesToGDP	-0.285				
	(0.567)				
ΔCommodityPerformance $\times$	1.095***				
ReservesToGDP	(0.399)				
KOF		0.0188			
		(0.0251)			
ΔCommodityPerformance $\times$		0.0126*			
KOF		(0.00651)			
ChinnIto			0.0400		
			(0.0256)		
ΔCommodityPerformance $\times$			0.0420		
ChinnIto			(0.0477)		
IBRDLoans				10.05*	
				(5.585)	
ΔCommodityPerformance $\times$				3.917	
IBRDLoans				(2.855)	
NetAidGNI					-0.0772**
					(0.0316)
ΔCommodityPerformance $\times$					0.0438
NetAidGNI					(0.0589)
Observations	129,316	129,316	127,087	124,170	117,413
R-squared	0.217	0.217	0.215	0.213	0.210
Number of Countries	34	34	33	34	34
Time & Country FE	Yes	Yes	Yes	Yes	Yes
Controls	Yes	Yes	Yes	Yes	Yes

This table shows results from OLS-panel regressions of the daily difference of a country's Emerging Market Bond Index Spread (ΔEMBI) on the daily returns on the weighted price index of a country's exported commodities (ΔCommodityPerformance) and controls. Interaction terms of ΔCommodityPerformance with ReservesToGDP ((1), official reserve assets in U.S. Dollar to GDP in U.S. Dollar), KOF ((2), KOF globalization index by Gygli et al. (2019)), ChinnIto ((3), Chinn-Ito capital account openness index by Chinn & Ito (2006)), IBRDLoans ((4), outstanding International Bank for Reconstruction and Development and International Development Association loans to GDP) and NetAidGNI ((5), net official development assistance to GNI) are estimated. Estimation period is from 01/01/1994 to 12/31/2016. Variable definitions are provided in Table A.3.3. Control variables include a country's stock index and exchange rate (to U.S. Dollar) returns, changes in the VIX, U.S. term spread, U.S. corporate spread, 10-year U.S. treasury yield and global government bond index. Estimations include country and time fixed effects on the monthly level. Standard errors (in parentheses) are clustered at the country level, ***, ** and * indicate statistical significance at the 1%, 5% and 10% level, respectively.

Figure 3.4: Marginal effects of commodity performance on EMBI spread returns interacted with policy measures against commodity dependence (2). Bars indicate 95% confidence intervals. Distribution of interaction variable is shown. Results of the corresponding regressions are in Table 3.9.

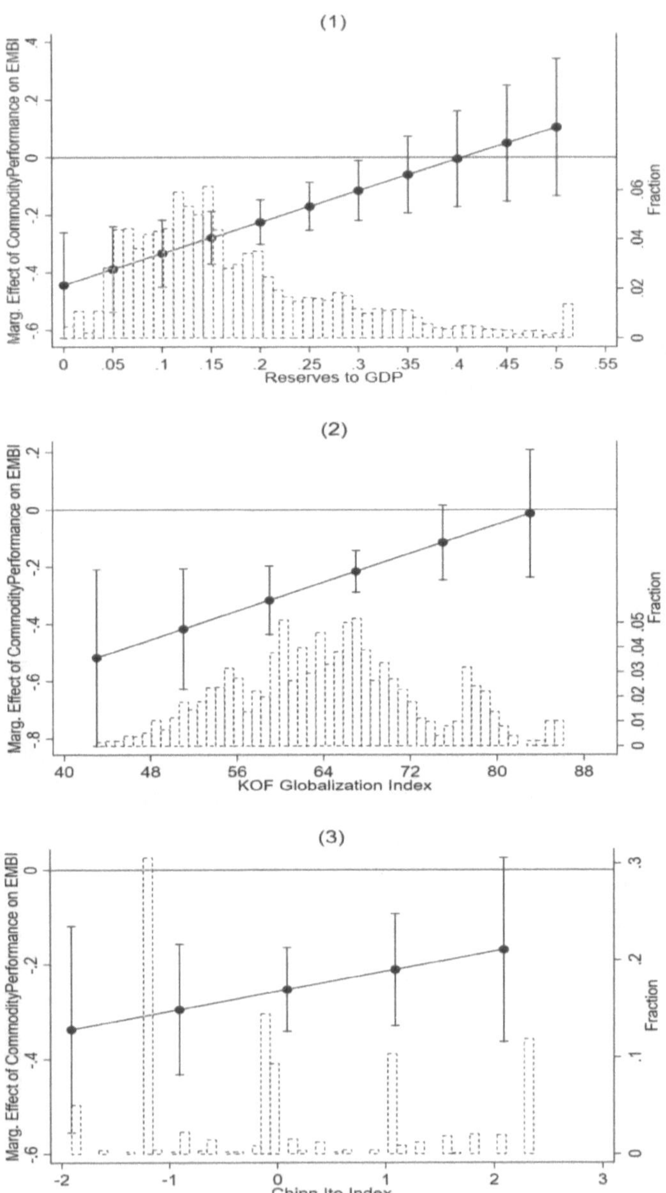

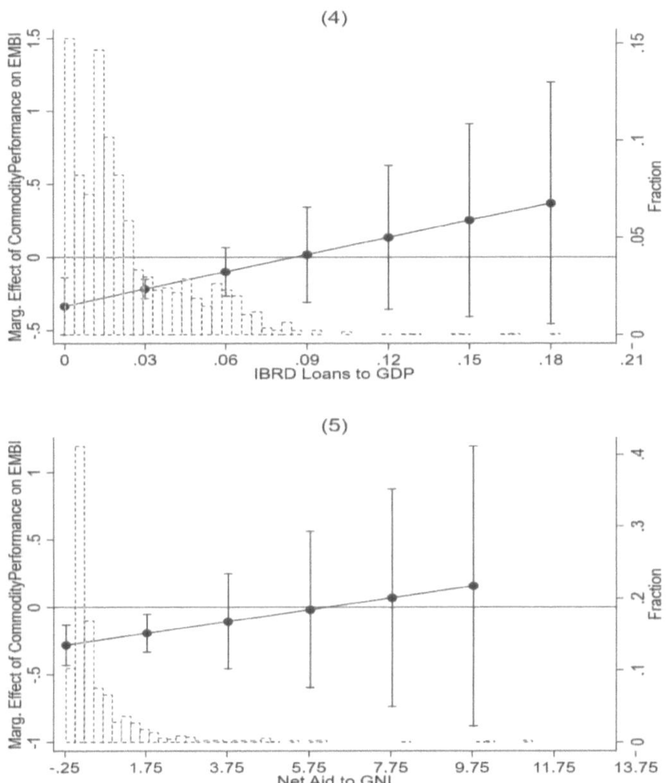

3.5 Robustness Checks

3.5.1 Dropping Countries with Liquidity Issues

We perform a range of sensitivity analyses to demonstrate the robustness of our results. First, we want to make sure that our results are not driven by liquidity issues some emerging markets might have in their EMBI or stock market data. To do so, we first drop all countries from the sample if their EMBI index turned temporarily illiquid during our estimation period, which we define formally as a constant EMBI spread for at least seven consecutive trading days. So far, we handled these periods by setting the affected EMBI changes to missing for the respective countries. Dropping the eight affected countries (Bulgaria, Croatia, Ivory

Coast, Morocco, Nigeria, Pakistan, Thailand, Tunisia) and repeating our baseline estimation (9) shows that the coefficient of export-weighted commodity price shocks is of similar size and statistical significance as in our main specification (Table 3.10, column (1)).

We then exclude the five countries (Ecuador, Ghana, Ivory Coast, Jamaica, Panama) whose stock market data has temporarily been varying only on a weekly instead of the daily level, which we previously handled the same way as with the EMBI returns. Results in column (2) report a commodity performance coefficient that is statistically significant at the 1% level, comparable to our main results.

Lastly, we drop all countries from the baseline estimation if they have less than 3000 business days of both stock market and EMBI return data which is somewhat over twelve years of data. This criterion affects 13 countries (Bulgaria, Croatia, Ghana, Ivory Coast, Jamaica, Kazakhstan, Panama, Serbia, Sri Lanka, Thailand, Tunisia, Ukraine, Vietnam). Results for this specification in column (3) yield a commodity performance coefficient that is extremely close to our main specification and statistically significant at the 1% level. These robustness checks indicate that our way of handling periods of lower liquidity in EMBI or stock data, by setting the respective data to missing if zero returns occur for at least two business days, was already a thorough method to deal with this issue and that any biases from low liquidity periods are limited in importance.

Table 3.10: Robustness: dropping countries with liquidity issues

	(1) ΔEMBI	(2) ΔEMBI	(3) ΔEMBI
ΔCommodityPerformance	-0.269***	-0.243***	-0.268***
	(0.0488)	(0.0462)	(0.0519)
Observations	108,062	116,645	100,043
R-squared	0.222	0.218	0.199
Number of Countries	26	29	21
Time & Country FE	Yes	Yes	Yes
Controls	Yes	Yes	Yes

This table shows results from OLS-panel regressions of the daily differences of a country's Emerging Market Bond Index Spread (ΔEMBI) on the daily returns on the weighted price index of a country's exported commodities (ΔCommodityPerformance) and controls. Robustness checks repeat baseline equation and include: (1): dropping all countries for which EMBI data turned, at some point, temporarily illiquid. (2): dropping all countries for which stock market data turned, at some point, temporarily illiquid. (3): dropping countries for which there are less than 3000 business days (roughly 12 years) of common EMBI and stock market data. Estimation period is from 01/01/1994 to 12/31/2016. Variable definitions are provided in Table A.3.3. Control variables include a country's stock index and exchange rate (to U.S. Dollar) returns, changes in the VIX, U.S. term spread, U.S. corporate spread, 10-year U.S. treasury yield and global government bond index. Estimations include country and time fixed effects on the monthly level. Standard errors (in parentheses) are clustered at the country level, ***, ** and * indicate statistical significance at the 1%, 5% and 10% level, respectively.

3.5.2 Alternative Specifications for EMBI and Commodity Performance

Next, we want to further alleviate concerns that our main results only hold because of the way we measured our variable of interest, i.e. export-weighted commodity price changes. We propose an alternative specification to capture price changes of key commodities of an emerging market that is similar to the procedure we used to take commodity imports into account in Section 3.3.2. We multiply price changes of a commodity with the absolute export value (in U.S. dollars) of the specific commodity for every country. We aggregate these value-weighted returns and then divide them by the GDP of each country. In this way, we take the importance of commodity exports on the share of a country's total economy into account, similarly as with the interaction model for the share of commodity exports on total exports. Our results are robust with respect to our previous findings, in that we report a coefficient of GDP-weighted commodity export returns that is statistically significant in affecting sovereign creditworthiness at the 1% level (Table 3.11, column (1)).

Though they are not part of the main GSCI index, there are additional GSCI spot price series for orange juice, palladium, platinum, bio-fuel, soybean oil and tin. We match these price series with the respective export volume of our sample countries and extend our commodity performance measure by these extra raw materials (except for orange juice which has no clear export match). Reassuringly, we find a slightly stronger commodity performance coefficient that is reported in column (2). However, the difference to the main specification is small, likely because these extra commodities otherwise would have been in the main index.

We already made sure that commodity prices cannot influence contemporaneous commodity weights by lagging all weights by one year. We take an additional step to make sure that export weights are unaffected by the price movement of the corresponding commodity by fixing each weight at its first observed value. For most countries, this first observation is in 1994. We hold this value constant for all periods thereafter. Our result is line with our baseline coefficient, and also statistically significant at the 1% level (column (3)), reassuring that most variation comes from commodity prices, whereas the weights are largely sticky.

In another robustness check we use sovereign credit default swap (CDS) premiums instead of EMBI spreads to measure sovereign risk. We draw CDS data from Thomson Reuters which is, however, only available since 2008 and only for 29 of our 34 sample countries. Comparable to our main specification, we take the first difference of the CDS spread as a new dependent variable. Even despite the data limitations, our commodity performance measure continues to have the expected negative sign and is statistically significant at the 1% level (column (4)).

Table 3.11: Robustness: alternative specifications for EMBI and commodity performance

	(1) ΔEMBI	(2) ΔEMBI	(3) ΔEMBI	(4) ΔCDS Spread
ΔCommodityPerformance: GDP-weighted	-2.422*** (0.562)			
ΔCommodityPerformance: additional commodities		-0.259*** (0.0434)		
ΔCommodityPerformance: constant weights			-0.250*** (0.0384)	
ΔCommodityPerformance				-0.455*** (0.0644)
Observations	127,357	129,901	129,901	54,982
R-squared	0.217	0.216	0.216	0.165
Number of Countries	34	34	34	29
Time & Country FE	Yes	Yes	Yes	Yes
Controls	Yes	Yes	Yes	Yes

This table shows results from OLS-panel regressions of the daily difference of a country's Emerging Market Bond Index Spread (ΔEMBI) on the daily returns on the weighted price index of a country's exported commodities (ΔCommodityPerformance) and controls. Robustness checks repeat baseline equation and include: (1): scaling CommodityPerformance to GDP. (2): including additional GSCI commodities in CommodityPerformance. (3): Holding the first observed commodity export weight constant. (4): using CDS spreads (first difference) as a dependent variable. Estimation period is from 01/01/1994 to 12/31/2016. Variable definitions are provided in Table A.3.3. Control variables include a country's stock index and exchange rate (to U.S. Dollar) returns, changes in the VIX, U.S. term spread, U.S. corporate spread, 10-year U.S. treasury yield and global government bond index. Estimations include country and time fixed effects on the monthly level. Standard errors (in parentheses) are clustered at the country level, ***, ** and * indicate statistical significance at the 1%, 5% and 10% level, respectively.

3.5.3 Alternative Control Variables

We further check that we have sufficiently controlled for any influences that could impact the relationship between commodity prices and sovereign risk. One further variation we might want to control for comes from credit risk in the U.S. interbank market that could spill over to emerging markets and which can be approximated by the TED spread. Though the TED spread enters with a positive sign and statistical significance when added to our main estimation, it does not change the significance level of our commodity performance measure which remains at the 1% level (Table 3.12, column (1)).

So far, we have not controlled for the economic performance of the U.S. Therefore, another potential variable worth including could be U.S. stock market returns, as they might affect both commodity prices and sovereign creditworthiness of emerging markets. However, adding the daily natural log returns of the S&P 500 to our main specification leaves the coefficient size and significance of the commodity performance almost unchanged (column (2)).

We also split up the term spread we included in our main specification and add changes in the 3-month U.S. treasury bill rate as an additional control variable. Our main results are not affected (column (3)).

Table 3.12: Robustness: alternative control variables

	(1) ΔEMBI	(2) ΔEMBI	(3) ΔEMBI
ΔCommodityPerformance	-0.278***	-0.233***	-0.249***
	(0.0467)	(0.0415)	(0.0437)
ΔTED-Spread	11.51***		
	(1.970)		
ΔS&P-500		-0.676***	
		(0.172)	
Δ3-Month-TBill-Yield			-0.0574
			(0.890)
Observations	115,634	129,316	129,316
R-squared	0.218	0.219	0.217
Number of Countries	34	34	34
Time & Country FE	Yes	Yes	Yes
Controls	Yes	Yes	Yes

This table shows results from OLS-panel regressions of the daily differences of a country's Emerging Market Bond Index Spread (ΔEMBI) on the daily returns on the weighted price index of a country's exported commodities (ΔCommodityPerformance) and controls. Robustness checks repeat baseline equation and include: (1): adding TED-Spread as additional control variable. (2): including S&P-500 return as additional control variable. (3): including 3-month-TBill-yield instead of TermSpread as additional control variable. Estimation period is from 01/01/1994 to 12/31/2016. Variable definitions are provided in Table A.3.3. Control variables include a country's stock index and exchange rate (to U.S. Dollar) returns, changes in the VIX, U.S. term spread, U.S. corporate spread, 10-year U.S. treasury yield and global government bond index. Estimations include country and time fixed effects on the monthly level. Standard errors (in parentheses) are clustered at the country level, ***, ** and * indicate statistical significance at the 1%, 5% and 10% level, respectively.

3.5.4 Alternative Fixed Effects, Frequency, Clustering and Time Series Results

In order to account for market-wide changes at a higher frequency, we replace the month fixed effects in our baseline estimation by week fixed effects. The commodity performance coefficient becomes just slightly smaller in size due to this procedure but remains statistically significant at the 1% level (Table 3.13, column (1)).

One further concern we want to alleviate is that the daily frequency in our data could be too noisy for a robust inference. We therefore collapse our data to the monthly frequency and repeat our baseline estimation. We obtain a somewhat higher coefficient of export-weighted commodity price shocks that is statistically significant at the 5% level (column (2)).

Furthermore, we cluster standard errors of our baseline both on the country and on the week level, to also allow for the correlation of errors within weeks. Our results remain statistically significant at the 1% level with this procedure (column (3)).

Next, we include day-of-the-week fixed effects to make sure our results are not driven by trading anomalies on certain business days, e.g. Fridays. Our main results are not affected by this approach (column (4)). We therefore conclude that the daily data structure is unlikely to be too noisy or biased with respect to our research design, but rather captures the maximum variation and information in the data.

Finally, we use time series regressions for each country to check for country-specific differences in the commodity-sovereign risk-dependence (Table 3.14). Overall, 26 out of the 34 sample countries have the expected negative sign of commodity performance in explaining sovereign risk. The coefficients are strongest for Argentina, Brazil, Ivory Coast, Kazakhstan, Peru, Philippines and Venezuela, which is reasonable, as these countries are in general large commodity exporters. Positive coefficients, though they are never statistically significant, are obtained e.g. for Poland or Hungary, which rely to lesser extends on commodity exports.

Table 3.13: Robustness: alternative fixed effects, frequency and clustering

	(1) ΔEMBI	(2) ΔEMBI	(3) ΔEMBI	(4) ΔEMBI
ΔCommodityPerformance	-0.209***	-0.433***	-0.249***	-0.249***
	(0.0421)	(0.205)	(0.0541)	(0.0441)
Observations	129,316	6,264	129,316	129,316
R-squared	0.256	0.285	0.217	0.217
Number of Countries	34	34	34	34
Week Time & Country FE	Yes	No	No	No
Month Time & Country FE	No	No	Yes	Yes
Year Time & Country FE	No	Yes	No	No
DayOfWeek FE	No	No	No	Yes
Controls	Yes	Yes	Yes	Yes
Cluster	Country	Country	Country & Week	Country

This table shows results from OLS-panel regressions of the daily differences of a country's Emerging Market Bond Index Spread (ΔEMBI) on the daily returns on the weighted price index of a country's exported commodities (ΔCommodityPerformance) and controls. Robustness checks repeat baseline equation and include: (1): applying weekly instead of monthly fixed effects. (2): collapsing data at monthly frequency. (3): clustering on country- and week-level. (4): including day-of-the-week fixed effects. Estimation period is from 01/01/1994 to 12/31/2016. Variable definitions are provided in Table A.3.3. Control variables include a country's stock index and exchange rate (to U.S. Dollar) returns, changes in the VIX, U.S. term spread, U.S. corporate spread, 10-year U.S. treasury yield and global government bond index. Estimations include country and time fixed effects on the monthly level. Standard errors (in parentheses) are clustered at the country level, ***, ** and * indicate statistical significance at the 1%, 5% and 10% level, respectively.

Table 3.14: Robustness: time series results of baseline estimation for each panel country

	(1) ΔEMBI Argentina	(2) ΔEMBI Brazil	(3) ΔEMBI Bulgaria	(4) ΔEMBI Chile	(5) ΔEMBI China	(6) ΔEMBI Colombia	(7) ΔEMBI Croatia	(8) ΔEMBI Ecuador	(9) ΔEMBI Egypt	(10) ΔEMBI Ghana	(11) ΔEMBI Hungary
ΔCommodity Performance	-0.600***	-0.640***	-0.208**	-0.0467	-0.124*	-0.404***	-0.408***	-0.421***	0.0208	0.0749	0.0315
	(0.168)	(0.136)	(0.0991)	(0.0494)	(0.0747)	(0.0755)	(0.133)	(0.116)	(0.0859)	(0.259)	(0.0974)
Observations	5,975	5,833	2,229	4,518	5,796	5,142	2,373	4,746	3,910	1,665	4,641
R-squared	0.202	0.220	0.400	0.168	0.176	0.380	0.326	0.155	0.382	0.352	0.244
Month Time & Country FE	Yes	Yes	Yes	Yes	Yes	Yes	Yes	Yes	Yes	Yes	Yes
Controls	Yes	Yes	Yes	Yes	Yes	Yes	Yes	Yes	Yes	Yes	Yes

This table shows results from OLS-panel regressions of the daily differences of a country's Emerging Market Bond Index Spread (ΔEMBI) on the daily returns on the weighted price index of a country's exported commodities (ΔCommodityPerformance) and controls. The robustness check repeats the baseline equation (Table 3.4) for each panel country as a time series regression. Estimation period is from 01/01/1994 to 12/31/2016. Variable definitions are provided in Table A.3.3. Control variables include a country's stock index and exchange rate (to U.S. Dollar) returns, changes in the VIX, U.S. term spread, U.S. corporate spread, 10-year U.S. treasury yield and global government bond index. Estimations include country and time fixed effects on the monthly level. Standard errors (in parentheses) are clustered at the country level, ***, ** and * indicate statistical significance at the 1%, 5% and 10% level, respectively.

Robustness: time series results of baseline estimation for each panel country (continued)

	(12) ΔEMBI Indonesia	(13) ΔEMBI Ivory Coast	(14) ΔEMBI Jamaica	(15) ΔEMBI Kazakhstan	(16) ΔEMBI Malaysia	(17) ΔEMBI Mexico	(18) ΔEMBI Morocco	(19) ΔEMBI Nigeria	(20) ΔEMBI Pakistan	(21) ΔEMBI Panama	(22) ΔEMBI Peru
ΔCommodity Performance	-0.282***	-0.757***	0.105	-0.516***	-0.0960*	-0.316***	-0.326**	0.0145	-0.0359	-0.265**	-0.544***
	(0.0946)	(0.219)	(0.135)	(0.117)	(0.0497)	(0.0636)	(0.154)	(0.120)	(0.0953)	(0.117)	(0.103)
Observations	3,224	1,936	2,186	2,436	5,131	5,971	3,348	3,656	3,242	2,136	5,116
R-squared	0.502	0.375	0.532	0.433	0.204	0.277	0.094	0.261	0.325	0.651	0.292
Month Time & Country FE	Yes	Yes	Yes	Yes	Yes	Yes	Yes	Yes	Yes	Yes	Yes
Controls	Yes	Yes	Yes	Yes	Yes	Yes	Yes	Yes	Yes	Yes	Yes

This table shows results from OLS-panel regressions of the daily differences of a country's Emerging Market Bond Index Spread (ΔEMBI) on the daily returns on the weighted price index of a country's exported commodities (ΔCommodityPerformance) and controls. The robustness check repeats the baseline equation (Table 3.4) for each panel country as a time series regression. Estimation period is from 01/01/1994 to 12/31/2016. Variable definitions are provided in Table A.3.3. Control variables include a country's stock index and exchange rate (to U.S. Dollar) returns, changes in the VIX, U.S. term spread, U.S. corporate spread, 10-year U.S. treasury yield and global government bond index. Estimations include country and time fixed effects on the monthly level. Standard errors (in parentheses) are clustered at the country level, ***, ** and * indicate statistical significance at the 1%, 5% and 10% level, respectively.

Robustness: time series results of baseline estimation for each panel country (continued)

	(23) ΔEMBI Philippines	(24) ΔEMBI Poland	(25) ΔEMBI Russia	(26) ΔEMBI Serbia	(27) ΔEMBI South Africa	(28) ΔEMBI Sri Lanka	(29) ΔEMBI Thailand	(30) ΔEMBI Tunisia	(31) ΔEMBI Turkey	(32) ΔEMBI Ukraine	(33) ΔEMBI Venezuela	(34) ΔEMBI Vietnam
ΔCommodity Performance	-0.469***	0.0197	-0.295***	0.105	-0.289***	0.0158	-0.102	-0.0602	-0.424***	-0.134	-0.561***	-0.167
	(0.0981)	(0.0951)	(0.0883)	(0.131)	(0.103)	(0.130)	(0.140)	(0.0856)	(0.117)	(0.162)	(0.111)	(0.113)
Observations	4,889	5,681	4,928	2,229	4,149	2,328	2,209	2,254	5,315	2,715	4,826	2,573
R-squared	0.464	0.254	0.267	0.389	0.460	0.375	0.121	0.260	0.359	0.247	0.177	0.441
Month Time & Country FE	Yes	Yes	Yes	Yes	Yes	Yes	Yes	Yes	Yes	Yes	Yes	Yes
Controls	Yes	Yes	Yes	Yes	Yes	Yes	Yes	Yes	Yes	Yes	Yes	Yes

This table shows results from OLS-panel regressions of the daily differences of a country's Emerging Market Bond Index Spread (ΔEMBI) on the daily returns on the weighted price index of a country's exported commodities (ΔCommodityPerformance) and controls. The robustness check repeats the baseline equation (Table 3.4) for each panel country as a time series regression. Estimation period is from 01/01/1994 to 12/31/2016. Variable definitions are provided in Table A.3.3. Control variables include a country's stock index and exchange rate (to U.S. Dollar) returns, changes in the VIX, U.S. term spread, U.S. corporate spread, 10-year U.S. treasury yield and global government bond index. Estimations include country and time fixed effects on the monthly level. Standard errors (in parentheses) are clustered at the country level, ***, ** and * indicate statistical significance at the 1%, 5% and 10% level, respectively.

3.6 Conclusion

This paper investigates the economic importance of commodity dependence on emerging markets' sovereign creditworthiness and derives macroeconomic and policy conditions that could propagate or curb this dependence. Using daily data for 34 emerging market economies from January 1, 1994 to December 31, 2016, we measure dependence as the impact of country-specific, export-weighted commodity price changes on EMBI sovereign bond yield spreads relative to U.S. Treasuries (controlling for a large set of major national and international financial indicators, country and time fixed effects). We obtain a statistically robust and economically meaningful commodity-sovereign risk dependence channel.

For the full set of countries, a one standard deviation increase in commodity price returns is associated on average with a 33.2 bps decrease in the EMBI spread differences. For heavy exporters (with a commodity export share on total exports equal or above the 90th percentile), the standardized effect yields a 47.5 bps decrease in EMBI spread differences which corresponds to 5% of the EMBI's standard deviation, and compares to around 40% of the standardized effects of the VIX and the corporate risk spread. Thus, particularly for commodity-dependent countries, commodity price fluctuations are a major determinant of sovereign creditworthiness. This average effect can be further differentiated along the characteristics of a country's commodity portfolio, national and international macroeconomic conditions and set of policy measures that affect commodity extraction.

We find, first, that commodity dependence increases with a larger share of commodity exports on total exportations. Exporting predominantly energy is associated with tighter commodity sensitivity, while the export of industrial metals with lower commodity dependence. Diversification within the commodity portfolio, i.e. being less concentrated on a single commodity, or exporting less price volatile commodities, however, does not seem to be associated with lower commodity dependence. As our later results show, a country can likely do more to tackle commodity dependence if it diversifies its economic structure towards downstream production and manufacturing sectors, instead of an additional commodity.

Second, we present evidence that commodity dependence increases in times of recession and lower public and private revenue streams. We do not find evidence that the outbreak of sovereign debt crises affect commodity dependence. Still, financial markets seem to pay attention to more recent incidents of sovereign defaults, which are associated with stronger commodity dependence the lower the number of years they date back. We also obtain strong evidence that more expansionary U.S. monetary policy spins the commodity cycle and increases commodity dependence significantly.

Third, we show consistent evidence that improving institutional quality can be a promising way to mitigate commodity dependence. All of our interactions variables, i.e. control of corruption, rule of law, political stability but also the progressiveness of a country's tax system, indicate that institutional quality is a decisive factor to tackle the dependence of a country's creditworthiness on raw material prices. We argue that better institutions likely increase transparency, provide clear ownership rights and limit corruption in the extraction process. The result also holds when focusing only on heavy commodity-exporting countries.

We also present results indicating that attracting more FDI flows, having larger manufacturing sectors and investing more in physical capital like machinery or infrastructure can be fruitful ways to reduce commodity dependence by fostering downstream production technologies. In contrast, having a low stock of foreign exchange reserves is associated with increasing commodity dependence. We also uncover that more open trade and financial accounts are associated with a weaker reliance on raw material prices. Lastly, development assistance measures such as received aid or World Bank loans show only a small effect for reducing commodity dependence.

A.3   Appendix to Chapter 3

Table A.3.1: Commodities towards which certain countries have world-market power

Commodity	Countries with world-market power
Cocoa	Brazil, Ghana, Indonesia, Ivory Coast, Malaysia, Nigeria
Coffee	Brazil, Colombia, Mexico, Vietnam
Corn	Argentina, Brazil, China, Ukraine
Cotton	China
Soybeans	Argentina, Brazil
Sugar	Brazil, Thailand
Wheat	Argentina, Russia
Lean Hogs	China
Cattle	
Energy	Russia
Aluminum	Russia
Copper	Chile, Indonesia, Peru
Lead	China, Mexico, Peru
Nickel	Russia
Zinc	China, Peru
Gold	
Silver	China, Mexico

This table shows which countries in the panel have world-market power in a certain commodity, as discussed in Section 3.3.2 with results reported in Table 3.5. For this test, commodities towards which a country has world-market power are removed from the commodity portfolio. World-market power is defined as having a global export share in a commodity of at least 10%. Meeting this criterion at least once at any point in time during our sample period leads to a world-market power classification.

Table A.3.2: Drivers of commodity-sovereign risk dependence: additional specifications

	(1) ΔEMBI	(2) ΔEMBI	(3) ΔEMBI	(4) ΔEMBI	(5) ΔEMBI
ΔCommodityPerformance	-0.545***	-0.688***	-0.424***	-0.989***	-0.833*
	(0.168)	(0.152)	(0.145)	(0.313)	(0.462)
ControlOfCorruption	-0.00649				
	(0.00479)				
ΔCommodityPerformance $\times$	0.00793**				
ControlOfCorruption	(0.00378)				
RuleOfLaw		-0.000529			
		(0.00475)			
ΔCommodityPerformance $\times$		0.0122***			
RuleOfLaw		(0.00307)			
PoliticalStability			-0.00387		
			(0.00940)		
ΔCommodityPerformance $\times$			0.00555		
PoliticalStability			(0.00460)		
KOF (de facto)				0.00413	
				(0.0194)	
ΔCommodityPerformance $\times$				0.0121**	
KOF (de facto)				(0.00483)	
KOF (de jure)					0.0217
					(0.0147)
ΔCommodityPerformance $\times$					0.00870
KOF (de jure)					(0.00672)
Observations	74,073	74,073	74,073	129,316	129,316
R-squared	0.232	0.233	0.231	0.217	0.217
Number of Countries	34	34	34	34	34
Time & Country FE	Yes	Yes	Yes	Yes	Yes
Controls	Yes	Yes	Yes	Yes	Yes

This table shows results from OLS-panel regressions of the daily difference of a country's Emerging Market Bond Index Spread (ΔEMBI) on the daily returns on the weighted price index of a country's exported commodities (ΔCommodityPerformance) and controls. The interactions with ControlOfCorruption (1), RuleOfLaw (2) and PoliticalStability (3) of Table 3.8 are repeated, however, they are specified for a subgroup of countries that have a commodity export share of more than 10% on total exports. In columns (4) and (5) the KOF globalization index of Table 3.9 is disentangled into the de facto and the de jure part. Estimation period is from 01/01/1994 to 12/31/2016. Variable definitions are provided in Table A.3.3. Control variables include a country's stock index and exchange rate (to U.S. Dollar) returns, changes in the VIX, U.S. term spread, U.S. corporate spread, 10-year U.S. treasury yield and global government bond index. Estimations include country and time fixed effects on the monthly level. Standard errors (in parentheses) are clustered at the country level, ***, ** and * indicate statistical significance at the 1%, 5% and 10% level, respectively.

Table A.3.3: Description and sources of variables

Variable	Description	Source
Variables in Baseline Regression (Section 3.3.1) (all variables are winsorized at 1st and 99th percentile, unless otherwise stated)		
ΔEMBI	Daily difference in Emerging Market Bond Index Spread (Global) (winsorized at 5th and 95th percentile)	J.P. Morgan

ΔCommodity Performance	Lagged-export-share weighted commodity price changes as described in Section 3.2.2	UN Comtrade, ITC Trade Map, S&P
ΔStock Returns	Daily change in natural logarithm of a country's general stock market index in U.S. Dollar	Datastream,[31] MSCI, S&P
ΔExchange Rate Returns	Daily percentage change of a country's local currency exchange rate towards the U.S. Dollar	Thomson Reuters
ΔVIX	Daily change in VIX volatility index	CBOE
ΔCorporate Spread	Daily change in spread between the S&P U.S. high yield corporate bond index and the corresponding investment grade index	S&P
Δ10-Year Treasury Yield	Daily change in the yield of 10-year U.S. Treasury bond	Datastream
ΔTerm Spread	Daily change in spread between 10-year U.S. Treasury yield and 3-month U.S. T-Bill yield	Datastream, Federal Reserve
ΔGlobal Government Bond Index	Daily change in natural logarithm of Bank of America Merrill Lynch Global Government Index	Merrill Lynch

Variables in Alternative Specification Regressions (Section 3.3.2)

ΔCommodity Performance: Excluding world-market -relevant exporters	CommodityPerformance excluding a commodity if country had at any time world-export share of more than 10%	
ΔCommodity Performance: Adjusting for imports	CommodityPerformance using net commodity exports (see Section 3.3.2)	
ΔCommodity Performance: Net Shock Indicator	CommodityPerformance using dummies for relevant commodities and extreme price events (see Section 3.3.2)	

Variables in Interaction Regressions (Section 3.4)
(all variables winsorized at 1st and 99th percentile, except indices)

Commodity Export Share	Share of commodity exports on total exports	UN Comtrade, ITC Trade Map
Commodity Standard Deviation	Rolling standard deviation of CommodityPerformance over past 23 business days (one month)	
Commodity HHI	Hirschman-Herfindahl-index (HHI), i.e. sum of squared export weights within CommodityPerformance	
Specialization: Agriculture, Energy, Industrial Metals, Precious Metals	1 if respective subgroup has share on total exports of at least 10%, 0 otherwise.	
GDP in local currency	GDP in constant prices, seasonally adjusted, local currency	Oxford Economics
GDP in U.S. Dollar	GDP in current prices and U.S. Dollar	World Bank
GDP Growth	Quarterly natural log growth rate of local currency GDP	
Tax Revenues to GDP	Government tax revenue (% of GDP, linearly interpolated)	World Bank
Corporate Profits to GDP	Corporate sector profits (% of GDP)	Oxford Economics
Debt to GDP	General gross government debt (% of GDP)	Oxford Economics

---

[31]For Ecuador, we merge local Quito Stock Exchange (in $) and S&P data to receive maximum coverage.

Inflation	Annual consumer price increase (winsorized at 5th and 95th percentiles to rule out hyperinflation periods)	World Bank
Sovereign Debt Crisis	Yearly dummy for sovereign debt crisis	Laeven & Valencia (2018)
Years since last Sovereign Debt Restructuring	Number of years since last sovereign debt restructuring. Value of 40 if no sovereign debt restructuring took place	Laeven & Valencia (2018)
GDP per Capita	GDP per capita in constant prices and U.S. Dollar	World Bank
Federal Funds Rate	U.S. federal funds effective rate	Federal Reserve
Control of Corruption	Control of corruption rank (The extent to which public power is exercised for private gain, including both petty and grand forms of corruption, as well as "capture" of the state by elites and private interests; linearly interpolated)	World Bank
Rule of Law	Rule of law rank (The extend of which agents have confidence in and abide by the rules of society; linearly interplt.)	World Bank
Political Stability	Political stability rank (The likelihood that the government will be destabilized by unconstitutional or violent means, including terrorism; linearly interpolated)	World Bank
Gini Redistribution	Absolute income redistribution (market-income inequality minus net-income inequality)	Solt (2019)
Manufacturing Share	Manufacturing value-added (% of GDP)	World Bank
FDI-Inflows	Net foreign direct investment inflows (% of GDP)	World Bank
GFCF	Gross Fixed Capital Formation (% of GDP)	World Bank
Reserves to GDP	Total reserve assets (% of GDP)	IMF
KOF	KOF Globalisation Index (composite index measuring globalization along several criteria such as trade and financial flows and regulation)	Gygli et al. (2019)
KOF (de facto/de jure)	KOF Globalisation Index disentangled into de facto and de jure component of index	Gygli et al. (2019)
Chinn-Ito	KAOPEN index of Chinn & Ito (2006) (index measuring regulatory controls over current or capital account transactions and exchange rate regimes)	Chinn & Ito (2006)
IBRD Loans	Outstanding loans from International Bank for Reconstruction and Development and International Development Association (% of GDP)	World Bank
Aid to GNI	Net official development assistance received (% of GNI; for Hungary, Bulgaria, Poland and Russia: % of GDP)	World Bank

Variables in Robustness Regressions (Section 3.5) (all variables winsorized at 1st and 99th percentile)		
ΔCDS Spread	Daily difference CDS spread (winsorized at 5th and 95h percentile)	Thomson Reuters
ΔTED Spread	Spread between 3-Month LIBOR based on U.S. dollars and 3-Month Treasury Bill	Fed St. Louis
ΔS&P 500	Daily change in natural log of Standard and Poor's 500 Composite	S&P
Δ3-Month T-Bill Yield	U.S. treasury bill 3-month yield	Federal Reserve

References to Chapter 3

Ahmed, S. & Zlate, A. (2014), 'Capital flows to emerging market economies: A brave new world?', *Journal of International Money and Finance* **48**, 221–248.

Aizenman, J., Edwards, S. & Riera-Crichton, D. (2012), 'Adjustment patterns to commodity terms of trade shocks: The role of exchange rate and international reserves policies', *Journal of International Money and Finance* **31**(8), 1990–2016.

Aizenman, J., Hutchison, M. & Jinjarak, Y. (2013), 'What is the risk of European sovereign debt defaults? Fiscal space, CDS spreads and market pricing of risk', *Journal of International Money and Finance* **34**, 37–59.

Aizenman, J., Jinjarak, Y. & Park, D. (2016), 'Fundamentals and sovereign risk of emerging markets', *Pacific Economic Review* **21**(2), 151–177.

Akram, Q. F. (2009), 'Commodity prices, interest rates and the dollar', *Energy economics* **31**(6), 838–851.

Alexeev, M. & Conrad, R. (2009), 'The elusive curse of oil', *The Review of Economics and Statistics* **91**(3), 586–598.

Arezki, R. & Brückner, M. (2011*a*), 'Oil rents, corruption, and state stability: Evidence from panel data regressions', *European Economic Review* **55**(7), 955–963.

Arezki, R. & Brückner, M. (2011*b*), 'Resource windfalls and emerging market sovereign bond spreads: The role of political institutions', *The World Bank Economic Review* **26**(1), 78–99.

Arezki, R. & Brückner, M. (2012), 'Commodity windfalls, democracy and external debt', *The Economic Journal* **122**(561), 848–866.

Bhattacharyya, S. & Hodler, R. (2010), 'Natural resources, democracy and corruption', *European Economic Review* **54**(4), 608–621.

Bouri, E., de Boyrie, M. E. & Pavlova, I. (2017), 'Volatility transmission from commodity markets to sovereign CDS spreads in emerging and frontier countries', *International Review of Financial Analysis* **49**, 155–165.

Bouri, E., Shahzad, S. J. H., Raza, N. & Roubaud, D. (2018), 'Oil volatility and sovereign risk of BRICS', *Energy Economics* **70**, 258–269.

Brambor, T., Clark, W. R. & Golder, M. (2006), 'Understanding interaction models: Improving empirical analyses', *Political analysis* pp. 63–82.

Bräuning, F. & Ivashina, V. (2020), 'US monetary policy and emerging market credit cycles', *Journal of Monetary Economics* **112**, 57–76.

Bun, M. J. & Harrison, T. D. (2019), 'OLS and IV estimation of regression models including endogenous interaction terms', *Econometric Reviews* **38**(7), 814–827.

Cashin, P., Céspedes, L. F. & Sahay, R. (2004), 'Commodity currencies and the real exchange rate', *Journal of Development Economics* **75**(1), 239–268.

Chen, Y.-c. & Rogoff, K. (2003), 'Commodity currencies', *Journal of International Economics* **60**(1), 133–160.

Chinn, M. D. & Ito, H. (2006), 'What matters for financial development? Capital controls, institutions, and interactions', *Journal of Development Economics* **81**(1), 163–192.

Clements, K. W. & Fry, R. (2008), 'Commodity currencies and currency commodities', *Resources Policy* **33**(2), 55–73.

Davis, G. A. (1995), 'Learning to love the dutch disease: Evidence from the mineral economies', *World Development* **23**(10), 1765–1779.

Deaton, A. (1999), 'Commodity prices and growth in Africa', *Journal of Economic Perspectives* **13**(3), 23–40.

Drechsel, T. & Tenreyro, S. (2018), 'Commodity booms and busts in emerging economies', *Journal of International Economics* **112**, 200–218.

Dreher, A., Lang, V. & Richert, K. (2019), 'The political economy of international finance corporation lending', *Journal of Development Economics* .

Fernández, A., González, A. & Rodriguez, D. (2018), 'Sharing a ride on the commodities roller coaster: Common factors in business cycles of emerging economies', *Journal of International Economics* **111**, 99–121.

Fernández, A., Schmitt-Grohé, S. & Uribe, M. (2017), 'World shocks, world prices, and business cycles: An empirical investigation', *Journal of International Economics* **108**, S2–S14.

Frankel, J. A. (2006), 'The effect of monetary policy on real commodity prices'. NBER Working Paper No. 12713.

Frankel, J. A. (2010), 'The natural resource curse: a survey'. NBER Working Paper No. 15836.

Gygli, S., Haelg, F., Potrafke, N. & Sturm, J.-E. (2019), 'The KOF globalisation index–revisited', *The Review of International Organizations* **14**(3), 543–574.

Gylfason, T. (2001), 'Natural resources, education, and economic development', *European Economic Review* **45**(4-6), 847–859.

Hartmann, P., Straetmans, S. & Vries, C. d. (2004), 'Asset market linkages in crisis periods', *Review of Economics and Statistics* **86**(1), 313–326.

Hilscher, J. & Nosbusch, Y. (2010), 'Determinants of sovereign risk: Macroeconomic fundamentals and the pricing of sovereign debt', *Review of Finance* **14**(2), 235–262.

James, A. (2015), 'The resource curse: A statistical mirage?', *Journal of Development Economics* **114**, 55–63.

Kohlscheen, E., Avalos, F., Schrimpf, A. et al. (2017), 'When the walk is not random: Commodity prices and exchange rates', *International Journal of Central Banking* **13**(2), 121–158.

Laeven, L. & Valencia, F. (2018), 'Systemic banking crises revisited'. IMF Working Paper 18/206.

Lang, V. F. & Tavares, M. M. M. (2018), 'The distribution of gains from globalization'. IMF Working Paper 18/54.

Longstaff, F. A., Pan, J., Pedersen, L. H. & Singleton, K. J. (2011), 'How sovereign is sovereign credit risk?', *American Economic Journal: Macroeconomics* **3**(2), 75–103.

Mehlum, H., Moene, K. & Torvik, R. (2006), 'Institutions and the resource curse', *The Economic Journal* **116**(508), 1–20.

Nizalova, O. Y. & Murtazashvili, I. (2016), 'Exogenous treatment and endogenous factors: Vanishing of omitted variable bias on the interaction term', *Journal of Econometric Methods* **5**(1), 71–77.

Reinhart, C. M., Reinhart, V. & Trebesch, C. (2016), 'Global cycles: Capital flows, commodities, and sovereign defaults, 1815-2015', *American Economic Review:: Papers & Proceedings* **106**(5), 574–80.

Reinhart, C. M., Rogoff, K. S. & Savastano, M. A. (2003), 'Debt intolerance'. NBER Working Paper No. 9908.

Sachs, J. D. & Warner, A. (1995), 'Economic reform and the process of global integration', *Brookings Papers on Economic Activity* **1995**(1), 1–118.

Sachs, J. D. & Warner, A. M. (2001), 'The curse of natural resources', *European Economic Review* **45**(4-6), 827–838.

Solt, F. (2019), 'Measuring income inequality across countries and over time: The standardized world income inequality database', *SWIID Version* **8**.

Temesvary, J., Ongena, S. & Owen, A. L. (2018), 'A global lending channel unplugged? Does US monetary policy affect cross-border and affiliate lending by global US banks?', *Journal of International Economics* **112**, 50–69.

The Economist (2017), 'Commodities are not always bad for you'. `https://www.economist.com/special-report/2017/10/05/commodities-are-not-always-bad-for-you`.

UNCTAD (2019), 'The state of commodity dependence 2019'. United Nations conference on trade and development.

World Economic Forum (2019), 'We must help developing countries escape commodity dependence'. `https://www.weforum.org/agenda/2019/05/why-commodity-dependence-is-bad-news-for-all-of-us/`.

Chapter 4:

Financial Linkages and Sectoral Business Cycle
Synchronization: Evidence from Europe

4.1 Motivation

Countries with stronger economic, cultural and political ties tend to have more synchronized output fluctuations. However, whether financial integration is one of such synchronizing factors for international business cycles is unresolved in the literature. On the one hand, Kose et al. (2003), Imbs (2006) and Morgan et al. (2004) find a positive relationship be-tween financial integration and business cycle synchronization. On the other hand, results by Kalemli-Ozcan, Papaioannou & Perri (2013) and Kalemli-Ozcan, Papaioannou & Pey-dró (2013) suggest that a higher degree of financial integration entails diverging patterns of economic activity. Cesa-Bianchi et al. (2019) argue that the nature of the shock, common (negative effect) or country-specific (positive effect), is what matters for synchronization.

We contribute to this literature, first, by assessing the effect of financial integration on economic activity not only among country pairs but across a multilateral network of directly and indirectly linked countries. More specifically, we use a flexible spatial model recently developed by Blasques et al. (2016) that combines time-varying matrices of economic distances, reflecting financial linkages, with a dynamic parameter approach.[32] Our dynamic spatial

---

[32](Static) spatial models have recently become popular in the empirical finance literature, see, e.g., Tonzer (2015), Herskovic et al. (2018), and Denbee et al. (2018).

model takes endogenous feedback and third-country effects into account.[33] In this way, we retrieve a scalar intensity parameter that reflects the extent of positive or negative business cycle co-movement over time. This set-up allows us to compare the extent of spillovers during recessions and tranquil periods. A static spatial model, in comparison, would only reveal the net effect of positive and negative spillovers, potentially hiding the heterogeneity of gross spillover shocks. As a second contribution, we dismantle the business cycle into its main industrial sectors, similarly to Schnabel & Seckinger (2015). As different industries can be exposed by different extents to shocks transmitted through financial links, this decomposition can give further insights on the conflicting results above.[34]

The analysis is based on a sample of 10 European countries over the period from 1996 to 2017, for which gross domestic product (GDP) growth is dissected into the value-added by 10 industries. We find that financially more integrated countries tend to have on average positive business cycle synchronization. This finding means that shocks are transmitted across countries via their financial linkages resulting in positive co-movements of GDP growth. However, the effect depends crucially on time and industry. Positive synchronization effects of financial integration on GDP and industries with business-sensitive cycles such as industrial production or wholesale & retail trade are much larger during crisis periods. Hence, for our sample of European countries, we can confirm previous findings in the literature (Kalemli-Ozcan, Papaioannou & Perri (2013)) while providing a more nuanced view on spillover intensities and their composition over time. Other industries are subject to small positive synchronization effects which are, interestingly, almost constant over both recessions and normal times (e.g. agriculture, construction). Cycle synchronization of a few industries is not subject to any positive or negative spillover effect stemming from financial integration, suggesting that these industries (arts & entertainment, public administration, real estate) are relatively closed-off and hardly affected by integrated financial markets. Therefore, time, industry-specification and feedback effects matter for the finance-business-cycle nexus. The paper is structured as follows. In the next section, we describe the data and the empirical method. Results are presented in Section 4.3. The last section concludes the paper.

---

[33] Acemoglu et al. (2012), for example, emphasize the relevance of the network structure for spillover effects between sectors.

[34] International co-movement through firms in one country and their cross-border links are analyzed by, for example, Di Giovanni et al. (2018) and Kleinert et al. (2015). The role of trade among countries for business cycle synchronization is assessed byAbbott et al. (2008) or Arkolakis & Ramanarayanan (2009).

4.2 Empirical Strategy

4.2.1 Data

We proxy the degree of financial integration using data on direct bilateral cross-border claims of banks from the Bank for International Settlements (BIS).[35] Similar to Kalemli-Ozcan, Papaioannou & Perri (2013), Cesa-Bianchi et al. (2019) and many others, we make use of the locational banking statistics as they are well-suited for this task. Compared to the consolidated banking statistics, cross-border inter-office positions between banks of the same group are not netted out. Thus, the locational statistics deliver a clear picture on cross-border linkages with the potential of generating spillovers.[36] From a theoretical perspective, this feature is important to consider as the activities of global banks matter for the transmission of shocks and effects on synchronization: Shocks in the real sector in one country should result in lower synchronization if global banks redirect their lending to unaffected countries. Shocks in the financial sector of some countries would induce global banks to retrench more globally, which in turn increases co-movement (Kalemli-Ozcan, Papaioannou & Perri 2013).

For Europe, the BIS currently reports complete data for 10 countries since 1995.[37] Based on this data, we can span a sizable network of European countries.[38] A snapshot of the network can be seen in Figure 4.1 for 2017:Q4. The graph reveals that some countries are more strongly interlinked than others, as reflected by the width of the links. The network overall shows a dense degree of interconnections. The sample period on which the estimations are based extends from 1996 until 2017 such that we can trace out whether the financial crisis starting in 2007/08 changes spillover dynamics permanently, or whether synchronization declines again, as one could hypothesize following the findings by Kalemli-Ozcan, Papaioannou & Perri (2013) for the tranquil period before the financial crisis. In Tables 4.1 and 4.2, we show examples for the weighting matrix at different points in time.

---

[35] While financial integration can take different forms, Hoffmann et al. (2019) find that banking integration dominates equity market integration in the euro area with the latter being still limited in size.

[36] The locational statistics have also the advantage of a stable coverage. While inter-office positions might change faster between inter-related offices compared to other cross-border positions, the series are still sufficiently stable such that no concerns due to volatility emerge.

[37] Belgium, Denmark, Finland, France, Germany, Ireland, Netherlands, Sweden, Switzerland, UK.

[38] We focus on European countries to obtain consistent data on value-added at the sectoral level from Eurostat. We thoroughly checked the data to have the best possible coverage of European countries and over time. Given that we are interested in time-varying spillovers, we have chosen those countries that report cross-border positions to the BIS over a long time period.

Figure 4.1: Bilateral cross-border claims (based on BIS Locational Banking Statistics) of countries' banking systems for 2017:Q4

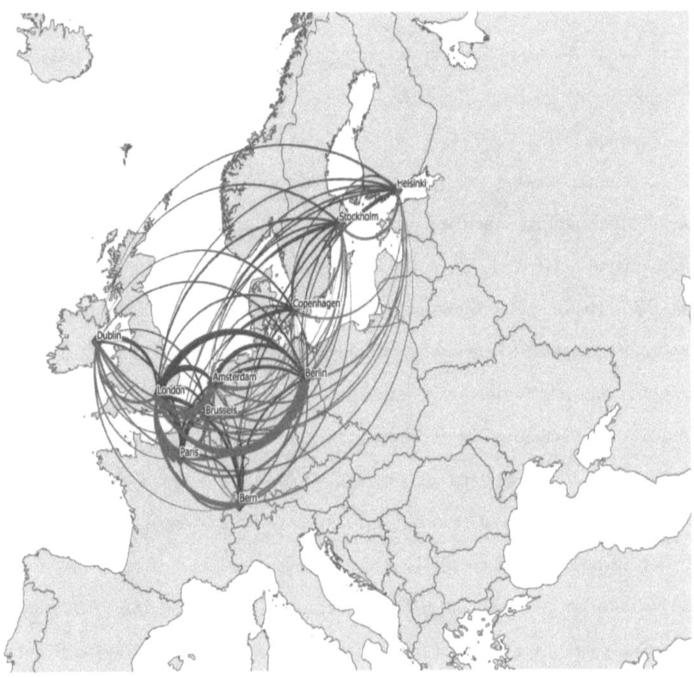

Table 4.1: Network matrix of banking systems' cross-border claims for 1996:Q1

	Belgium	Denmark	Finland	France	Germany	Ireland	Netherlands	Sweden	Switzerland	UK
Belgium		1374	1526	28657	10514	5974	14907	3855	5348	47685
Denmark	2038		1341	3136	1533	2533	850	5507	969	15507
Finland	617	470		786	372	17	316	3914	233	3916
France	29439	6380	2241		39688	3529	15750	4918	16766	146083
Germany	13509	5466	4576	28134		11273	21636	6820	13429	116563
Ireland	1126	650	794	2400	8621		1647	1474	479	12090
Netherlands	21963	2280	703	14368	21862	6041		3768	7602	40884
Sweden	1351	1594	1656	897	1746	277	2117		778	17531
Switzerland	36673	1475	927	31597	23261	1868	34426	2498		173198
UK	41960	9614	9224	96405	161291	17703	33092	25833	39088	

This table shows the network matrix based on BIS locational banking statistics for 1996:Q1. Data on a country's banking system claims (from row-country towards column-country) in millions of US dollars is depicted. In the empirical analysis, we use row-normalized versions of these matrices.

Table 4.2: Network matrix of banking systems' cross-border claims for 2017:Q4

	Belgium	Denmark	Finland	France	Germany	Ireland	Netherlands	Sweden	Switzerland	UK
Belgium		2423	1150	71463	52457	20012	72993	2444	14870	72486
Denmark	1998		7661	12851	47964	2114	3039	83375	3520	31498
Finland	1318	12614		5858	8170	496	2583	21551	143	8985
France	122957	10260	7133		118875	67037	108142	18416	76632	372761
Germany	36694	20771	22328	216641		34132	173043	41449	79548	322777
Ireland	6312	4006	908	19655	16158		26158	2445	2416	94892
Netherlands	61170	4268	10620	85493	81552			6724	29611	318205
Sweden	2930	104404	96840	12968	29685	1102	9286		5897	64214
Switzerland	9654	5510	1109	57596	49264	8158	23053	4374		148994
UK	56175	22008	17182	507496	439557	163186	315718	44461	199079	

This table shows the network matrix based on BIS locational banking statistics for 2017:Q4. Data on a country's banking system claims (from row-country towards column-country) in millions of US dollars is depicted. In the empirical analysis, we use row-normalized versions of these matrices.

To capture the business cycle in a country and as our main dependent variable, we use quarterly GDP growth in constant prices drawn from the OECD. We then decompose GDP into quarterly gross value-added growth, also in constant prices, of 10 major industries downloaded from Eurostat.[39] All growth rates are winsorized at the 1st and 99th percentile. Figure 4.2 shows the average pattern of GDP growth and sectoral growth over time. Obviously, a sharp decline can be detected for aggregate GDP growth as well as for most sectors following the financial crisis starting in 2007/08. The growth path of some sectors closely resembles the one of aggregate GDP growth (e.g., industry (except construction) or wholesale & retail trade) while some sectors have notably different dynamics.

---

[39]Agriculture, forestry and fishing. Arts, entertainment, recreation and other services. Construction. Financial and insurance activities. Industry (except construction). Information and communication. Professional, scientific and tech activities. Public administration, deference, education, human health and social work. Real estate activities. Wholesale and retail trade, transport, accommodation and food.

Figure 4.2: Quarterly growth rates of GDP and industrial sectors across panel countries (1996-2017)

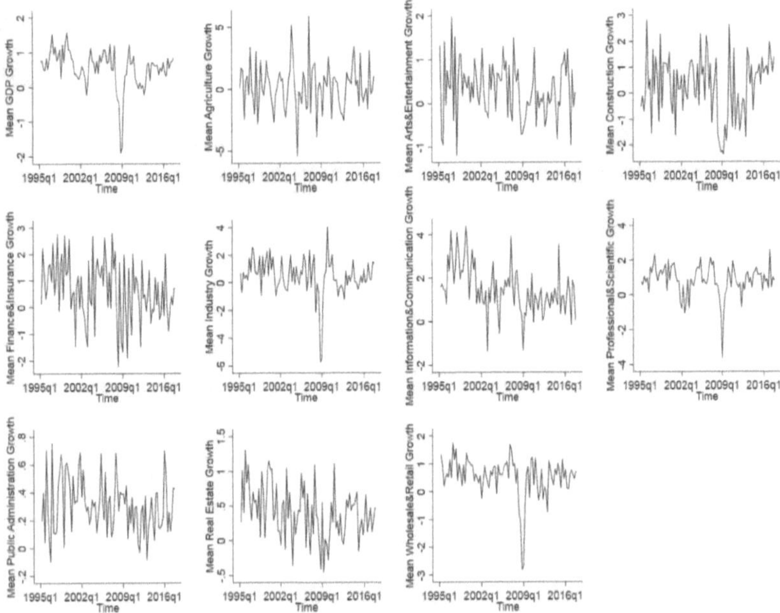

In Tables 4.3-4.5, we show that quarterly GDP growth rates across countries are correlated to different extents. However, we take a purely bilateral perspective in this case. In the estimations, we explicitly account for the fact that also indirect links can contribute to business cycle synchronization. Still, such simple descriptive statistics reveal important facts. On average, there is less evidence for negative co-movements. Supporting the findings by Kalemli-Ozcan, Papaioannou & Perri (2013), during the crisis period (Table 4.4), correlations go up, which does apply to most country pairs but excludes those with Ireland. Comparing the pre- and post-crisis periods, no general pattern emerges. Partially, correlations are lower, while for some other country pairs they are still at a higher level (Tables 4.3 and 4.5). In sum, these patterns support the idea to control for time-varying spillover dynamics instead of taking a static view.

Table 4.3: Correlation of pre-crisis GDP growth rates between countries (1996:Q1–2007:Q4)

	Belgium	Denmark	Finland	France	Germany	Ireland	Netherlands	Sweden	Switzerland	UK
Belgium	1.000									
Denmark	0.175	1.000								
Finland	0.214	-0.141	1.000							
France	0.432	0.303	0.315	1.000						
Germany	0.365	0.274	0.133	0.368	1.000					
Ireland	0.339	0.234	0.144	0.176	0.095	1.000				
Netherlands	0.425	0.424	0.136	0.507	0.448	0.218	1.000			
Sweden	0.222	0.121	0.141	0.411	0.094	0.074	0.348	1.000		
Switzerland	0.462	0.060	0.233	0.469	0.345	0.202	0.348	0.307	1.000	
UK	0.006	-0.064	-0.009	0.104	-0.007	0.214	0.082	-0.005	0.160	1.000

Table 4.4: Correlation of financial crisis GDP growth between countries (2008:Q1–2009:Q4)

	Belgium	Denmark	Finland	France	Germany	Ireland	Netherlands	Sweden	Switzerland	UK
Belgium	1.000									
Denmark	0.846	1.000								
Finland	0.830	0.743	1.000							
France	0.889	0.751	0.849	1.000						
Germany	0.918	0.752	0.875	0.991	1.000					
Ireland	0.152	-0.081	0.148	0.137	0.109	1.000				
Netherlands	0.721	0.546	0.811	0.860	0.873	-0.111	1.000			
Sweden	0.764	0.510	0.794	0.748	0.762	0.433	0.780	1.000		
Switzerland	0.925	0.775	0.874	0.860	0.912	0.005	0.811	0.761	1.000	
UK	0.922	0.680	0.685	0.891	0.907	0.139	0.684	0.619	0.805	1.000

Table 4.5: Correlation of post-crisis GDP growth between countries (2010:Q1–2016:Q4)

	Belgium	Denmark	Finland	France	Germany	Ireland	Netherlands	Sweden	Switzerland	UK
Belgium	1.000									
Denmark	0.369	1.000								
Finland	0.551	0.188	1.000							
France	0.399	0.018	0.423	1.000						
Germany	0.626	0.082	0.450	0.564	1.000					
Ireland	0.181	-0.318	-0.140	0.381	0.205	1.000				
Netherlands	0.338	0.026	0.446	0.441	0.327	0.265	1.000			
Sweden	0.062	-0.229	-0.038	-0.118	0.008	0.094	0.534	1.000		
Switzerland	0.088	0.072	0.136	0.187	0.302	0.170	0.134	-0.102	1.000	
UK	0.137	0.045	0.192	0.132	0.325	0.155	0.189	0.237	0.220	1.000

To get a first glimpse of which sectors correlate most closely with aggregate growth, we show in Table 4.6 correlations between GDP growth and industrial sector growth rates. In line with the graphical evidence (Figure 4.2), correlations are highest between GDP growth and the industry (except construction) as well as the wholesale & retail trade sectors. The lowest correlation emerges with the agricultural sector. These differences highlight that economies' aggregate growth paths can be determined by diverging sectoral developments such that taking a more granular and sectoral view can provide useful insights.

Table 4.6: Correlations between quarterly GDP and industrial sector growth rates

	ΔGDP	ΔAgriculture, Forestry and Fishing	ΔArts, Entertainment and Recreation	ΔConstruction	ΔFinancial and Insurance Activities	ΔIndustry	ΔInformation and Communication	ΔProfessional, Scientific and Tech Activities	ΔPublic Administration	ΔReal Estate Activities	ΔWholesale and Retail Trade, Transport
ΔGDP	1.000										
ΔAgriculture, Forestry and Fishing	0.133	1.000									
ΔArts, Entertainment and Recreation	0.150	-0.002	1.000								
ΔConstruction	0.385	-0.022	0.127	1.000							
ΔFinancial and Insurance Activities	0.264	0.086	0.053	0.053	1.000						
ΔIndustry	0.662	0.047	0.065	0.114	0.081	1.000					
ΔInformation and Communication	0.337	0.043	0.043	0.085	0.021	0.129	1.000				
ΔProfessional, Scientific and Tech Activities	0.347	0.048	0.091	0.144	0.044	0.168	0.097	1.000			
ΔPublic Administration	0.199	0.004	0.047	0.077	0.061	0.015	0.120	0.033	1.000		
ΔReal Estate Activities	0.232	0.013	0.019	0.069	-0.015	0.093	0.033	0.025	0.060	1.000	
ΔWholesale and Retail Trade, Transport	0.535	0.020	0.167	0.278	0.073	0.191	0.170	0.281	0.042	0.119	1.000

This table shows correlations between GDP and industrial growth rates for the period 1996–2017. All variables are in quarterly growth rates. GDP and industrial sector growth rates are winsorized at the 1st and 99th percentile. Data is obtained from the OECD and Eurostat. See Table A.4.12 for definitions and sources.

We require further national and global control variables that might affect the finance-business-cycle nexus. On the country-level, we include quarterly growth rates of labor productivity, consumer confidence, labor force, gross fixed capital formation, government expenditure, credit to the non-financial sector (in percent of GDP). Furthermore, we include the quarterly growth rate of exports of goods and services (in percent of GDP) as export reliance might be related to business cycle fluctuations. The role of global factors has, e.g., been analyzed by Kose et al. (2012) or Cerutti et al. (2019), and on the international level, we control for the quarterly change of the VIX and the Euro to U.S. Dollar exchange rate. More information on the variables can be found in Table A.4.12. Summary statistics on the dependent and explanatory variables can be found in Table 4.7. We provide a correlation table between the dependent variable and the controls in Table 4.8.

Table 4.7: Summary statistics of all variables

Variable	Obs.	Mean	Median	Std. Dev.	Min	Max
ΔGDP	910	0.553	0.532	0.897	-2.275	4.004
ΔAgriculture, Forestry and Fishing	906	0.178	0.128	5.205	-19.24	19.72
ΔArts, Entertainment and Recreation	906	0.320	0.295	1.756	-5.420	5.914
Δ Construction	906	0.279	0.296	2.468	-8.088	7.895
ΔFinancial and Insurance Activities	906	0.636	0.538	3.019	-9.018	10.46
ΔIndustry	906	0.445	0.429	2.699	-9.753	11.16
ΔInformation and Communication	898	1.488	1.236	2.667	-6.968	12.06
ΔProfessional, Scientific and Tech Activities	906	0.835	0.784	1.828	-4.825	6.882
ΔPublic Administration	906	0.312	0.305	0.624	-1.486	2.267
ΔReal Estate Activities	906	0.405	0.336	1.089	-2.785	3.901
ΔWholesale and Retail Trade, Transport	898	0.521	0.592	1.274	-4.337	4.053
Δ Productivity	916	0.313	0.270	1.197	-5.830	21.71
ΔCredit to Non-Financial Sector	910	0.482	0.374	1.912	-9.325	28.86
ΔLabourForce	910	0.202	0.179	0.533	-1.650	7.990
ΔConsumer Confidence	907	0.0159	0.0413	0.562	-2.327	2.274
ΔGFCF	910	0.867	0.674	7.326	-46.95	161.1
ΔGovernment Expenditure	910	0.410	0.388	0.926	-4.893	6.657
ΔExportsToGDP	910	0.392	0.509	2.784	-14.75	13.66
ΔEuro-to-Dollar Exchange Rate	910	0.000484	0.00282	0.0490	-0.116	0.118
ΔVIX	910	-0.00367	-0.0120	0.274	-0.664	1.052

Sample period is 1996:Q1–2017:Q4. All variables are in quarterly growth rates. GDP and industrial sector growth rates are winsorized at the 1st and 99th percentile. See Table A.4.12 for more information on data sources.

Table 4.8: Correlations between dependent and independent variables

	ΔGDP	ΔAgriculture, Forestry and Fishing	ΔArts, Entertainment and Recreation	ΔConstruction	ΔFinancial and Insurance Activities	ΔIndustry	ΔInformation and Communication	ΔProfessional, Scientific and Tech Activities	ΔPublic Administration	ΔReal Estate Activities	ΔWholesale and Retail Trade, Transport
ΔProductivity	0.739	0.032	0.101	0.209	0.216	0.590	0.282	0.180	0.070	0.115	0.352
ΔConsumer Confidence	0.229	0.057	-0.011	0.075	0.075	0.166	0.007	0.093	-0.034	-0.059	0.188
ΔLabourForce	0.110	-0.073	0.089	0.145	0.007	-0.020	0.027	0.069	0.078	0.058	0.110
ΔGFCF	0.198	0.012	0.084	0.239	0.022	0.068	0.010	0.032	0.002	0.009	0.124
ΔGovernment Expenditure	0.150	0.030	0.032	0.106	0.027	0.083	0.013	0.082	0.256	0.083	0.056
ΔCredit to Non-Financial Sector	-0.000	-0.052	0.049	-0.037	0.085	0.023	0.056	-0.058	0.030	-0.015	-0.066
ΔExportsToGDP	0.219	-0.009	-0.002	-0.004	0.018	0.218	0.068	0.167	-0.049	0.003	0.221
ΔEuro-to-Dollar Exchange Rate	-0.001	0.003	-0.041	0.076	0.002	-0.051	0.037	0.047	-0.026	-0.011	-0.036
ΔVIX	-0.112	0.026	-0.038	-0.029	-0.028	-0.127	0.041	-0.048	-0.002	-0.002	-0.058

This table shows correlations between dependent and independent variables for the period 1996–2017. All variables are in quarterly growth rates. GDP and industrial sector growth rates are winsorized at the 1st and 99th percentile. See the data description in Table A.4.12 for more information on data sources.

4.2.2 Method

Our empirical methodology comes from the literature on time-varying spatial dependence as established by Blasques et al. (2016). Compared to the related literature, we do not calculate bilateral correlations between countries' GDP growth and explain those correlations. Instead, we model each country's GDP growth as a weighted function of all financially interlinked countries' GDP growth. The spatial modeling approach has the advantage that interdependencies between a large set of countries can simultaneously be taken into account, and that the possibility of contemporaneous spillovers of shocks is incorporated.

Spatial models require the specification of a spatial weights matrix, which is typically chosen as a function of physical or economic distances between units. In our case, economic distance is defined by the cross-border bank claims two countries hold towards another, which is a measure of the degree of financial integration. We use a spatial lag model, which implies that each country's dependent variable may react to shocks to both the regressors and the disturbances of neighboring countries. Third-country and feedback effects are automatically taken into account. Additionally, we employ a time-varying spatial dependence parameter approach as suggested in Blasques et al. (2016). In this way, the magnitude of cross-sectional spillovers transmitted by financial integration can vary over time, allowing us to compare the effects during different stages of the economic and financial cycle.

The score-driven spatial lag model is given by

$$y_t = \rho_t W_t y_t + X_t \beta + e_t, \quad e_t \sim p_e(0; \Sigma, \nu), \quad t = 1, ..., T, \tag{11}$$

where y_t denotes an $N \times 1$- vector of country-specific growth rates of GDP or industrial value-added at time t. X_t is a matrix of country-specific and international regressors and $\beta = (\beta_1, ..., \beta_M)'$ is a vector of unknown coefficients. The dependent variable and the explanatory variables are demeaned to control for country fixed effects. We control for time-varying dynamics affecting all countries alike by including global controls. p_e denotes the density of the vector of disturbances e_t and Σ is a positive definite covariance matrix. We consider normally and Student's t-distributed disturbances. In the case of Student's t-distributed disturbances, p_e also depends on a degrees of freedom parameter ν.

The key ingredient of a spatial model is the term $\rho_t W_t y_t$. The matrix W_t reflects the degree of financial integration between countries at time t and is assumed to be weakly exogenous. Tables 4.1-4.2 provide examples of such a matrix for one point in time. The scalar spatial dependence parameter ρ_t measures the intensity of cross-country shock spillovers of real output y_t that are induced by financial links. To ensure stability, we specify $\rho_t = h(f_t)$ where $h(\cdot)$ is a monotone transformation such that $\rho_t \in (-1, 1)$. To describe the dynamics of f_t, we adopt the autoregressive score framework of Creal et al. (2011, 2013) and Harvey (2013).

The score framework centers around the use of the scaled score of the conditional density p_e to drive the time-variation in f_t. The updating equation for f_t is given by

$$f_{t+1} = \omega + As_t + Bf_t, \tag{12}$$

where ω, A, and B are fixed unknown parameters, and $s_t = S_t \nabla_t$ is the scaled score function, which serves as innovation term for the time-varying parameter.

The spatial dependence coefficient ρ_t may be interpreted as an indicator of the degree of business cycle synchronization driven by financial links: A positive coefficient would reveal evidence for business cycle synchronization, while a negative coefficient would point towards desynchronization. Importantly, the modeling approach takes into account that in highly integrated markets, business cycle synchronization does not only occur between two countries in isolation. For example, shocks to country A can spill over to the directly linked country B but also affect country C, which has in turn financial links to country B.

Our approach has the advantage that we can gain information on the strength of direct versus indirect (or third-country) spillovers. The econometric mechanism behind that is explained in the following. Repeatedly inserting the model for the vector of dependent variable observations gives rise to a power series expansion

$$y_t = X_t\beta + \rho_t W_t X_t\beta + \rho_t^2 W_t^2 X_t\beta + ... + e_t + \rho_t W_t e_t + \rho_t^2 W_t^2 e_t + ... \tag{13}$$

which is useful to disentangle "ordinary" impacts of shocks to regressors and disturbances of each unit (measured by the term $X_t\beta + e_t$) from "first-round" spillovers from direct neighbors $\rho_t W_t(X_t\beta + e_t)$ and "third-country" effects, which are contained in the higher-order terms.

Under stability, equation (13) converges to the nonlinear reduced form

$$y_t = Z_t X_t \beta + Z_t e_t, \tag{14}$$

with $Z_t = (I - \rho_t W_t)^{-1}$. Equation (14) shows that shocks to country i's regressors X_{it} or disturbance e_{it} can spill over to potentially many other countries' GDP growth, as Z_t is in general a full matrix. Feedback effects to country i itself occur if, for example, the weights w_{ij} and w_{ji} are non zero. If no spillovers are present, i.e. $\rho_t = 0$, the model reduces to the linear regression model

$$y_t = X_t \beta + e_t. \tag{15}$$

Instead of imposing a particular model specification ex ante, we determine empirically whether GDP growth rates and industrial value-added are indeed driven by shock spillovers from other countries. In particular, we estimate three versions of the model, each assuming either normally or Student's t distributed disturbances: (1) a baseline specification without any spillovers, i.e. $\rho_t = 0$, $t = 1, ..., T$, (2) a static version with $\rho_t = \rho$, $\forall t$ and (3) the dynamic specification given in equation (11). Model selection is conducted using the Akaike Information Criterion corrected for small sample sizes (AICc). This way, we are agnostic about whether spillovers are present or not as well as whether spillovers are constant or time-varying. By identifying the model that best fits the data, we gain insights into the nature of spillovers, which might bear important policy implications. For example, if spillovers are time-varying, spillover dynamics need to be monitored more closely and policy measures to curb them need to be flexible enough.

4.3   Estimation Results

4.3.1   Results for Overall Output Fluctuations (GDP)

Our formal approach tests which spatial model is the most appropriate to analyze business cycle propagation. In doing so, we address the hypothesis if business cycles between financially integrated countries can be considered as static and independent from another, or if there are dynamic spillovers present between countries. Table 4.9 shows the results using quarterly GDP growth as the dependent variable for different specifications of the spatial model. Columns (1) and (2) report results obtained using a model without spillovers, columns (3) and (4) show findings allowing for spillovers with a static dependence coefficient, and columns (5) and (6) display results for a model with time-varying spillover effects of financial integration on GDP growth. The errors are assumed to be either normally or t-distributed. Values of the AICc indicate that the data favor the model using time-varying spillover effects and a t-distribution with the AICc being lower by 83 points than the no spillover model and by 16 points compared to the static model. Allowing for time-varying spillovers as introduced in Section 4.2.2 therefore seems to be the most appropriate way to measure the effect of financial integration on output fluctuations.

Turning to the coefficients, we find strong evidence for spatial dependence both in the static and the dynamic model, with ρ as well as A and B, the parameters entering the dynamic updating equation (12), being large and statistically highly significant. This result thus supports that, from a regional perspective, business cycles should not be considered in isolation in financially integrated countries as dynamics can propagate via financial links towards a country's own GDP growth.

All other coefficients enter with largely expected signs in all models. For example, productivity growth enters with a positive and economically and statistically highly significant coefficient. In our preferred specification, that features dynamic spillovers and t-distributed errors (column (6)), we obtain that rising consumer confidence, labor force, government expenditure, export growth, and a depreciating Euro towards the U.S. Dollar are positively and statistically significantly associated with higher GDP growth. Gross fixed capital formation growth (with a positive coefficient) and credit ratios (with a negative coefficient) also have a significant impact, but only on a 10% level. Changes in the volatility index VIX enter with a positive sign, which is, however, statistically insignificant.[40]

Having established that dynamic spillover effects from financial integration matter for European countries' GDP growth, we now turn to the evolution of the spillover intensity parameter over time. The spillover parameter varies between minus and plus one which represents diverging or respectively converging impacts of financial integration on business cycle spillovers. Depicting the parameter over time, as in Figure 4.3, allows us to address the research question if, when and by how much financial integration induces business cycle synchronization. We observe a strong cyclical component in Figure 4.3. While the parameter

---

[40]The results are robust and similar towards employing an alternative model specification using OLS regressions with country and year fixed effects, see Table A.4.1.

Table 4.9: Baseline results: gross domestic product

	no spillovers		static spillovers		dynamic spillovers	
	(1) normal ΔGDP	(2) t ΔGDP	(3) normal ΔGDP	(4) t ΔGDP	(5) normal ΔGDP	(6) t ΔGDP
ρ			0.404***	0.294***		
			(0.027)	(0.032)		
ω					0.079***	0.033
					(0.021)	(0.029)
A					0.047***	0.043***
					(0.005)	(0.006)
B					0.721***	0.823***
					(0.066)	(0.072)
$\ln(\sigma^2)$	-0.699***	-1.408***	-0.938***	-1.422***	-0.955***	-1.439***
	(0.048)	(0.095)	(0.048)	(0.088)	(0.047)	(0.086)
constant	-0.192***	-0.231***	-0.153***	-0.234***	-0.146***	-0.234***
	(0.028)	(0.025)	(0.025)	(0.023)	(0.025)	(0.022)
ΔVIX	0.054	0.188**	0.030	0.133**	0.040	0.093
	(0.089)	(0.079)	(0.079)	(0.070)	(0.075)	(0.063)
ΔEuroToDollar	-0.370**	0.956	-0.063	0.486	0.812**	1.001***
	(0.490)	(0.405)	(0.435)	(0.379)	(0.399)	(0.348)
ΔProductivity	0.494***	0.717***	0.438***	0.679***	0.434***	0.669***
	(0.021)	(0.025)	(0.019)	(0.025)	(0.019)	(0.025)
ΔCreditToNonFinancialSector	-0.113***	-0.011	-0.093***	-0.017	-0.094***	-0.018
	(0.013)	(0.011)	(0.012)	(0.011)	(0.009)	(0.011)
ΔConsumerConfidence	0.229***	0.144***	0.103**	0.091**	0.118***	0.108***
	(0.045)	(0.038)	(0.041)	(0.037)	(0.040)	(0.034)
ΔGFCF	0.010***	0.005*	0.007**	0.005*	0.007**	0.005*
	(0.003)	(0.003)	(0.003)	(0.003)	(0.003)	(0.003)
ΔLabourForce	0.144***	0.151***	0.098**	0.128***	0.089**	0.116***
	(0.045)	(0.037)	(0.040)	(0.036)	(0.039)	(0.035)
ΔGovernmentExpenditure	0.069***	0.066***	0.064***	0.062***	0.067***	0.056***
	(0.026)	(0.020)	(0.023)	(0.019)	(0.023)	(0.019)
ΔExportToGDP	0.070***	0.037***	0.040***	0.031***	0.031***	0.027***
	(0.009)	(0.007)	(0.008)	(0.007)	(0.008)	(0.007)
ν		4.209***		5.428***		5.697***
		(0.760)		(1.071)		(1.147)
Observations	880	880	880	880	880	880
logLik	-941.016	-800.064	-850.217	-765.314	-836.895	-754.370
AICc	1907.506	1628.289	1728.594	1561.546	1707.544	**1545.406**

This table shows estimation results for different model specifications (no spillovers, static spillovers, dynamic spillovers) based on normal or t-distribution as indicated in the column header. Standard errors are reported in brackets below coefficient estimates. The Akaike information criterion (AICc) is depicted in bold for the model with the best fit (smallest value). The dependent variable is quarterly GDP growth. The sample period spans 1996–2017. All variables are in quarterly growth rates. GDP and industrial sector growth rates are winsorized at the 1st and 99th percentile. ***, ** and * indicate statistical significance at the 1%, 5% and 10% level, respectively. See Table A.4.12 for more information on data sources.

is positive on average, it peaks during times of financial stress, in particular the dotcom
bubble (around 2000), the financial crisis (around 2008) and the European debt crisis (2011-
2013). In calmer times (mid-1990s or mid-2000s), the spillover strength is lower, whereas
in the recent time period (mid-2010s) characterized by financial disintegration, the spillover
strength even becomes negative. Despite our focus on a sample of European countries, this
result is broadly in line with Kalemli-Ozcan, Papaioannou & Perri (2013) who identify that
financial integration has stronger effects on business cycle synchronization in times of crisis.
By tracing the spillover intensity over the sample period, we contribute to the literature by
showing in a more granular way time-varying dynamics of spillovers. For example, we can
detect whether spillover intensity slowly trends upward or quickly peaks. From Figure 4.3,
it can be seen that movements, especially in times when spillover intensity increases, tend to
occur relatively fast suggesting that policy instruments that are aligned to the cycle and can
be quickly activated might work best.

Figure 4.3: Time-varying spillover strength for gross domestic product

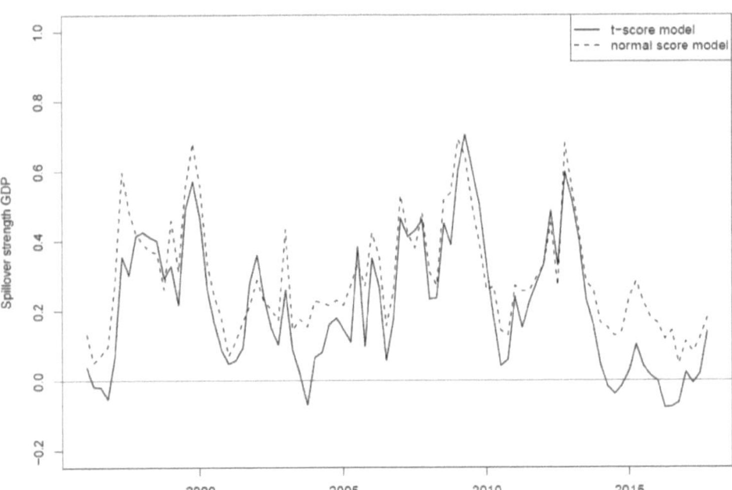

This graph shows the time-varying spillover strength for the model with quarterly GDP growth as the dependent variable and the sample period 1996-2017. The corresponding estimation results are in Table 4.9.
The model is estimated based on a score-driven model with Student's t-distributed (solid line) and normally
distributed (dashed line) disturbances.

As laid out in Section 4.2.2, the score-driven spatial model can be used to disentangle spillover effects from shocks that occur because two countries are directly linked, and shocks that are due to second- or higher-order effects. For example, consider an isolated unit shock to the disturbance of the first unit in vector y_t, i.e. suppose $e_t = (1, 0 \cdots 0)'$ in equation (11).[41] Equations (13) and (14) show that apart from the unit impact that the shock has on country 1's GDP growth itself, there are spillovers to its first-degree "neighbors", which are contained in the first column of $\rho_t W_t$, and, additionally, third-country feedback effects that can be summarized in the first column of the matrix

$$(I_N - \rho_t W_t)^{-1} - I_N - \rho_t W_t. \tag{16}$$

We report the average magnitude of such direct and third-country spillover effects in Table 4.10. That is, we compute in how far a shock to one of the sample countries would lead to "direct" spillovers to all other countries with which it maintains financial linkages, and compute the average effect across these interlinked countries. Furthermore, we compute the average "third-country" effect, that arises due to indirect linkages between the shocked country and all other countries. The results show that there is substantial heterogeneity across countries as concerns the relative sizes of the average direct and indirect effects. For example, spillover effects are the largest following a shock to the United Kingdom reflecting its role as a financial center. In contrast, shocks in Finland and Switzerland have reduced direct and third-country effects. Yet there are also countries which induce relatively large direct effects but third-country effects are small such as Sweden, which might reflect that Sweden maintains strong links with neighboring countries but plays a minor role for indirect spillovers. In sum, the results show that only considering direct shock spillovers affecting output correlation leads to an underestimation of the full spillover that arises due to third-country effects.

---

[41]For simplicity, we abstract from the regressor term here.

Table 4.10: Direct and third-country spillover effects

	direct	third-country
Belgium	0.1042	0.0334
Denmark	0.1257	0.0237
Finland	0.0747	0.0173
France	0.2337	0.0854
Germany	0.2990	0.1009
Ireland	0.1053	0.0409
Netherlands	0.1926	0.0660
Sweden	0.1751	0.0304
Switzerland	0.0819	0.0402
UK	0.7818	0.1657

This table reports the average total spillover effects after separate unit shocks to each country's disturbance. Column "direct" corresponds to the column sums of matrix $\bar{\rho}W$ where a bar denotes the sample average. Column "third-country" corresponds to the column sums of the sample average of the matrix in equation (16).

4.3.2 Results for Industrial Output Fluctuations

We now apply the analysis to individual industrial sectors. This dissection can shed light on the components that drive the aggregated effects on GDP we discussed in the previous section. Note that certain sectors are highly sensitive towards the business cycle, such as industrial output or wholesale and retail trade. For these sectors, we therefore expect similar patterns, both in terms of the best-fitting model and for the graph of the time-varying spillovers, compared to GDP growth. For other industrial sectors, we do not expect the same sensitivity towards growth shocks from other countries. An example is public administration, a largely nationally determined sector that should not vary much with business cycle spillovers due to international financial integration.

As before, we estimate six versions of each model (no spillovers, constant and dynamic dependence parameter, and both normal and t-errors), but for brevity, we only show the respective best model in Table 4.11 according to the AICc.[42] The results in Table 4.11 can be summarized as follows. Time-varying spillover dynamics fit the data best for four out of ten industrial sectors: industry (excluding construction); information & communication; professional, scientific and tech activities; wholesale & retail trade. Three different sectors show the best fit for a model that allows for spillovers, which are, however, driven by a constant instead of a time-varying parameter: agriculture, forestry and fishing; construction; financial and insurance activities. Finally, there are three cases, arts and entertainment, public administration and real estate, for which we do not find evidence for any spatial dependence.

---

[42]The full tables covering all models per industrial sector be seen in the Appendix, Table A.4.2 and following.

Figure 4.4 depicts the time-varying dependence parameters for the four industries, for which the time-varying model turned out to be the best fit.

The patterns of three industrial sectors' spillover parameters are similar to the graph for GDP, see Figure 4.3. The resemblance is most obvious for the wholesale & retail trade sector, but it is also present for industry (excluding construction) as well as for professional, scientific and tech activities.[43] These three industries have in common that they strongly depend on the current business cycle, making them similarly affected towards spillovers from financial integration as overall GDP. Hence, both our estimation results and the graphs suggest that interlinked countries specializing in these sectors are very likely to face a higher extent of business cycle synchronization, which relates to the results by Imbs (2004). Our findings also suggest that in these industries, cross-border links of banks play a prominent role by creating a network in which shocks are transmitted for the same sector across countries.[44] One reasoning can be that within the above mentioned sectors the banking sector is channeling more funds because there is a larger presence of multinational firms resulting in cross-country investments.[45] A further explanation can go back to the fact that some sectors rely more on external finance and might thus also depend more on cross-border capital flows.[46]

We also observe strong, time-varying spillovers across European countries' IT sectors (information & communication). However, the pattern in the upper right panel of Figure 4.4 differs from the ones for the other three sectors. Our interpretation of this finding is that IT sectors do face growth spillovers driven by financial interconnections, but in a more idiosyncratic way, potentially reflecting the more volatile and disruptive nature of this sector.

There is no evidence for spatial dependence for the public administration sector. One likely explanation is that reliance on cross-border capital is lower in this sector, which in turn implies a lower degree of financial integration and thus potential links to transmit shocks.

---

[43]This sector comprises mostly legal, management or engineering activities,

[44]The role of shock spillovers between interconnected sectors depending on the network structure is discussed by Acemoglu et al. (2012).

[45]The presence of multinational firms as such can result in international co-movements (Cravino & Levchenko (2017), Di Giovanni et al. (2018), Kleinert et al. (2015)). Such co-movements can be fueled in case multinational banks follow their multinational customers (Buch (2000)), in this context, it is also important that we do not net out inter-office positions between banks.

[46]E.g., for the United States, Laeven & Valencia (2013) show high values of external dependence for machinery, other industries or professional goods. Kroszner et al. (2007) find that sectors relying more on external finance via banks are hit more by banking crises and consequently tighter financial constraints (see also, Braun & Larrain (2005), Chava & Purnanandam (2011)).

Table 4.11: Industrial sector results (best model fit)

	(1) ΔAgriculture	(2) ΔArts & Entertainment	(3) ΔConstruction	(4) ΔFinance & Insurance	(5) ΔIndustry	(6) ΔInformation & Communication	(7) ΔProfessional, Scientific & Tech Activities	(8) ΔPublic Administration	(9) ΔReal Estate	(10) ΔWholesale & Retail Trade
ρ	0.181***		0.228***	0.174***						
	(0.050)		(0.043)	(0.047)						
ω					0.043	0.111	0.121**			0.049
					(0.040)	(0.089)	(0.061)			(0.046)
A					0.023***	0.043**	0.043***			0.036**
					(0.007)	(0.019)	(0.013)			(0.017)
B					0.747***	0.543*	0.546***			0.713***
					(0.205)	(0.327)	(0.174)			(0.217)
$\ln(sigma^2)$	-0.488***	-0.372***	-0.324***	-0.21***	-0.894***	-0.437***	-0.386***	-0.249***	-0.187**	-0.565***
	(0.094)	(0.086)	(0.072)	(0.071)	(0.080)	(0.082)	(0.074)	(0.074)	(0.073)	(0.078)
constant	-0.003	-0.073**	-0.057	-0.044	-0.107***	-0.107***	0.007	-0.146***	-0.068*	-0.052
	(0.036)	(0.036)	(0.037)	(0.038)	(0.030)	(0.038)	(0.036)	(0.038)	(0.039)	(0.033)
ΔVIX	0.095	-0.098	-0.005	-0.016	-0.088	0.081	0.038	-0.079	0.083	0.172*
	(0.100)	(0.103)	(0.115)	(0.120)	(0.081)	(0.104)	(0.109)	(0.115)	(0.124)	(0.101)
ΔEuroToDollar	-0.483	-1.076*	1.339**	0.103	-0.579	0.889	0.489	-0.88	-0.098	0.224
	(0.557)	(0.609)	(0.619)	(0.655)	(0.456)	(0.599)	(0.585)	(0.618)	(0.656)	(0.556)
ΔProductivity	0.079**	0.055*	0.129***	0.118***	0.423***	0.187***	0.103***	0.035	0.094***	0.201***
	(0.033)	(0.030)	(0.028)	(0.030)	(0.030)	(0.030)	(0.026)	(0.029)	(0.031)	(0.029)
ΔCreditTo NonFinancialSector	-0.012	-0.011	-0.045***	0.003	-0.03**	-0.004	-0.049***	-0.001	-0.015	-0.05***
	(0.017)	(0.017)	(0.016)	(0.017)	(0.014)	(0.017)	(0.016)	(0.018)	(0.018)	(0.016)
ΔConsumer Confidence	0.021	-0.029	0.106*	0.08	0.044	-0.101*	0.074	-0.026	-0.111*	0.128**
	(0.053)	(0.055)	(0.058)	(0.061)	(0.044)	(0.055)	(0.056)	(0.058)	(0.062)	(0.052)
ΔGFCF	0.001	0.007*	0.015***	-0.002	0	-0.005*	-0.002	-0.001	-0.002	0.002
	(0.003)	(0.004)	(0.004)	(0.004)	(0.003)	(0.003)	(0.004)	(0.004)	(0.004)	(0.003)
ΔLabourForce	-0.088	0.145**	0.136**	0.01	0.023	0.009	0.028	0.087	0.041	0.135***
	(0.056)	(0.061)	(0.059)	(0.058)	(0.046)	(0.059)	(0.054)	(0.063)	(0.061)	(0.051)
ΔGovernment Expenditure	-0.012	0.036	0.052	0.006	0.039	-0.012	0.05	0.282***	0.031	0.046
	(0.032)	(0.034)	(0.033)	(0.035)	(0.025)	(0.034)	(0.032)	(0.037)	(0.036)	(0.030)
ΔExportToGDP	-0.002	0.004	-0.01	0.004	0.049***	0.006	0.027**	-0.009	0.017	0.042***
	(0.012)	(0.011)	(0.012)	(0.012)	(0.010)	(0.012)	(0.011)	(0.012)	(0.012)	(0.011)
ν	4.927***	6.502***	12.481***	14.569***	7.437***	7.257***	12.978***	12.532***	13.456***	9.464***
	(1.019)	(1.485)	(3.805)	(5.171)	(1.673)	(1.655)	(4.447)	(4.112)	(4.644)	(2.634)
Observations	880	880	880	880	880	880	880	880	880	880
logLik	-1184.567	-1199.966	-1173.074	-1213.063	-960.234	-1165.225	-1147.117	-1202.135	-1225.42	-1085.497
AICc	2400.052	2428.092	2377.066	2457.044	1957.136	2367.117	2330.9	2432.43	2479.001	2207.661

This table shows estimation results for the different industrial sectors whereas the model specification with the best fit is shown. All estimates are based on a t-distribution. Standard errors are reported in brackets below coefficient estimates. The dependent variable is the quarterly growth of the sector as indicated in the column header. The sample period spans 1996–2017. All variables are in quarterly growth rates. GDP and industrial sector growth rates are winsorized at the 1st and 99th percentile. ***, ** and * indicate statistical significance at the 1%, 5% and 10% level, respectively. See the data description in Table A.4.12 for more information on data sources.

Similar results can be observed for the sector arts & entertainment and the real estate sector, which makes sense as both are largely nationally influenced. In the estimation results, these sectors showed the best fit for a no-spillover model, suggesting that they are largely unaffected by spillover effects stemming from financial market integration.

We observe positive but rather small spillover effects for the agricultural, construction and financial sectors. Furthermore, the data for all these sectors favor a model with a static spillover coefficient.[47] These results may indicate that the mentioned sectors are exposed to spillovers due to financial integration, but not as prime candidates and/or in a cyclical manner. From a policy perspective, this bears the implication that rescue measures during crisis times should be internationally coordinated, but also take the country-specific characteristics of each sector sufficiently into account.

Figure 4.4: Time-varying spillover strength for four industrial sectors

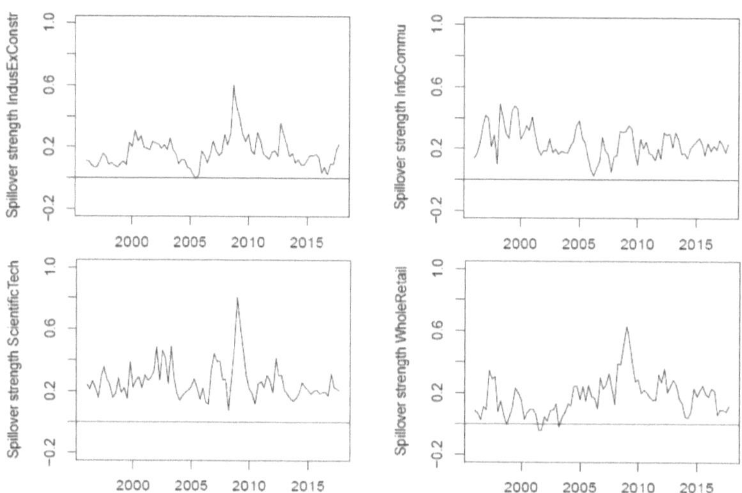

The graph shows the time-varying spillover strength for the four industrial sectors for which the time-varying model is the best. The dependent variable is the quarterly sectoral growth rate and the sample period spans 1996-2017. The corresponding estimation results are in Table 4.11. The sectors comprise: Upper left: Industry (except construction); Upper right: Information and Communication; Lower left: Professional, scientific and tech activities; Lower right: Wholesale and retail trade, transport, accommodation and food). The model is estimated based on score-driven model with Student's t-distributed disturbances.

---

[47] The time-varying spillover coefficients for these three sectors are almost constant throughout our estimation period as can be seen in the Appendix, Figure A.4.1.

4.4 Conclusion

Whether financial integration between countries leads to diverging or converging patterns of GDP growth is still not fully resolved in the literature. We shed new light on this issue by modeling a group of financially developed European countries as a financial network, thereby extending the pure bilateral framework used in the literature, and taking dynamic feedback effects within the network into account. We arrive at two major results.

First, spillover effects via the channel of financial integration on business cycles vary over time and are much stronger during periods of financial turmoil. Second, business-sensitive sectors like industrial production, wholesale & retail trade, or professional, scientific & tech activities are strongly exposed to spillover effects from financial integration, with time-variation following a similar cyclical pattern as for overall GDP. Industrial sectors such as agriculture, construction or finance also feature positive spillover effects, but are less affected, and the spillover intensity does not vary over time. Nationally influenced sectors such as public administration, arts & entertainment and real estate are not subject to relevant spillovers, positive or negative, due to financial integration.

Our results bear important policy implications. As we consistently find evidence for positive spillover dynamics across European countries and over time, our results show that in a densely financially integrated network of countries, business cycles are co-moving. Consequently, focusing only on national approaches to stabilize business cycles is likely to have limited effects, especially during times of crisis. In contrast, national measures should be accompanied by supranational actions mitigating spillovers of shocks via cross-border links among banking system, which supports policy measures such as the establishment of a European Banking Union. Furthermore, we show that industries are exposed to growth spillovers at different extents. This finding implies that in order to evaluate the exposure of a country's economy to business cycle synchronization and to mitigate negative effects during crisis times, sectoral specializations have to be taken into account. Only then can policy rescue programs be more effectively designed to support the sectors in distress.

A.4 Appendix to Chapter 4

Table A.4.1: OLS comparison regression to dynamic model in Table 4.9

	(1) ΔGDP
ΔVIX	-0.0738
	(0.0717)
ΔEuroToDollar	-0.495**
	(0.193)
ΔProductivity	0.518***
	(0.0486)
ΔCreditToNonFinancialSector	-0.0964**
	(0.0311)
ΔConsumerConfidence	0.106
	(0.0589)
ΔGFCF	0.00616
	(0.00370)
ΔLabourForce	0.161***
	(0.0477)
ΔGovernmentExpenditure	0.0896*
	(0.0412)
ΔExportToGDP	0.0125
	(0.00741)
Constant	0.323
	(0.0849)
Observations	907
Number of Countries	10
R-squared	0.712
Country FE	Yes
Year FE	Yes

This table shows estimation results of an OLS regression of the same variables as in the dynamic model in Table 4.9. Country and year fixed effects are included. Standard errors are reported in brackets below coefficient estimates. The dependent variable is quarterly GDP growth. The sample period spans 1996-2017. All variables are in quarterly growth rates. GDP and industrial sector growth rates are winsorized at the 1st and 99th percentile. ***, ** and * indicate statistical significance at the 1%, 5% and 10% level, respectively. See the data description in Table A.4.12 for more information on data sources.

Table A.4.2: Additional estimation results: agriculture, forestry and fishing

	no spillovers		static spillovers		dynamic spillovers	
	(1)	(2)	(3)	(4)	(5)	(6)
	normal	t	normal	t	normal	t
	ΔAgri-culture	ΔAgri-culture	ΔAgri-culture	ΔAgri-culture	ΔAgri-culture	ΔAgri-culture
ρ			0.142	0.181		
			(0.047)	(0.050)		
ω					0.022	0.182
					(0.000)	(0.115)
A					-0.034	0.032
						(0.030)
B					0.884	-0.035
					(0.003)	(0.569)
$\ln(\sigma^2)$	-0.019	-0.464	-0.032	-0.488	-0.054	-0.489
	(0.048)	(0.094)	(0.048)	(0.094)	(0.044)	(0.094)
Constant	0.025	-0.009	0.025	-0.003	0.023	0.000
	(0.040)	(0.037)	(0.040)	(0.036)	(0.040)	(0.037)
Δ VIX	0.089	0.092	0.093	0.095	0.057	0.089
	(0.126)	(0.103)	(0.125)	(0.100)	(0.106)	(0.098)
ΔEuroToDollar	-0.001	-0.482	0.003	-0.483	0.479	-0.465
	(0.688)	(0.572)	(0.683)	(0.557)	(0.690)	(0.549)
Δ Productivity	0.024	0.080	0.024	0.079	0.018	0.078
	(0.030)	(0.033)	(0.030)	(0.033)	(0.030)	(0.033)
ΔCreditToNonFinancialSector	-0.011	-0.012	-0.011	-0.012	-0.009	-0.013
	(0.018)	(0.017)	(0.018)	(0.017)	(0.018)	(0.017)
ΔConsumerConfidence	0.094	0.036	0.078	0.021	0.088	0.019
	(0.063)	(0.054)	(0.063)	(0.053)	(0.055)	(0.053)
ΔGFCF	0.001	0.001	0.001	0.001	0.001	0.001
	(0.005)	(0.003)	(0.005)	(0.003)	(0.004)	(0.003)
ΔLabourForce	-0.088	-0.088	-0.082	-0.088	-0.084	-0.092
	(0.063)	(0.057)	(0.063)	(0.056)	(0.061)	(0.056)
ΔGovernmentExpenditure	-0.022	-0.008	-0.023	-0.012	-0.021	-0.013
	(0.036)	(0.032)	(0.036)	(0.032)	(0.036)	(0.032)
ΔExportToGDP	-0.007	-0.003	-0.008	-0.002	-0.009	-0.003
	(0.012)	(0.012)	(0.012)	(0.011)	(0.012)	(0.011)
ν		4.961		4.927		4.926
		(1.027)		(1.019)		(1.016)
logLik	-1240.245	-1190.744	-1235.842	-1184.567	-1230.286	-1183.974
AICc	2505.963	2409.649	2499.845	**2400.052**	2494.326	2404.615

Notes: This table shows estimation results for different model specifications (no spillovers, static spillovers, dynamic spillovers) based on normal or t-distribution as indicated in the column header. Standard errors are reported in brackets below coefficient estimates. Empty standard errors imply that the numerical matrix of standard errors is not positive and that the corresponding spillover model is not well suited to estimate this industrial sector. The Akaike information criterion (AICc) is depicted in bold for the model with the best fit (smallest value). The dependent variable is quarterly growth of the sector "Agriculture, forestry and fishing". The sample period spans 1996–2017. All variables are in quarterly growth rates. GDP and industrial sector growth rates are winsorized at the 1st and 99th percentile. See the data description in Table A.4.12 for more information on data sources.

Table A.4.3: Additional estimation results: arts, entertainment, recreation and other services

	no spillovers		static spillovers		dynamic spillovers	
	(1)	(2)	(3)	(4)	(5)	(6)
	normal	t	normal	t	normal	t
	ΔArts & Entertainment	ΔArts & Entertainment	ΔArts & Entertainment	ΔArts & Entertainment	ΔArts & Entertainment	ΔArts & Entertainment
ρ			0.033	0.062		
			(0.051)	(0.051)		
ω					0.025	0.047
					(0.045)	(0.056)
A					0.029	0.020
					(0.023)	(0.026)
B					0.292	0.244
					(0.467)	(0.605)
$\ln(\sigma^2)$	-0.033	-0.372	-0.034	-0.378	-0.036	-0.377
	(0.048)	(0.086)	(0.048)	(0.086)	(0.048)	(0.086)
Constant	-0.057	-0.073	-0.055	-0.069	-0.048	-0.064
	(0.040)	(0.036)	(0.040)	(0.036)	(0.040)	(0.036)
Δ VIX	-0.070	-0.098	-0.071	-0.098	-0.066	-0.102
	(0.125)	(0.103)	(0.125)	(0.102)	(0.124)	(0.102)
ΔEuroToDollar	-1.141	-1.076	-1.111	-0.986	-1.061	-0.978
	(0.683)	(0.609)	(0.684)	(0.608)	(0.681)	(0.603)
Δ Productivity	0.078	0.055	0.077	0.054	0.075	0.053
	(0.030)	(0.030)	(0.030)	(0.030)	(0.030)	(0.030)
ΔCreditToNonFinancialSector	-0.007	-0.011	-0.007	-0.011	-0.004	-0.010
	(0.018)	(0.017)	(0.018)	(0.016)	(0.018)	(0.017)
ΔConsumerConfidence	-0.048	-0.029	-0.048	-0.030	-0.041	-0.028
	(0.063)	(0.055)	(0.063)	(0.055)	(0.063)	(0.055)
ΔGFCF	0.007	0.007	0.007	0.007	0.007	0.007
	(0.005)	(0.004)	(0.005)	(0.004)	(0.005)	(0.004)
ΔLabourForce	0.116	0.145	0.113	0.142	0.112	0.141
	(0.063)	(0.061)	(0.063)	(0.061)	(0.063)	(0.061)
ΔGovernmentExpenditure	0.011	0.036	0.011	0.036	0.012	0.036
	(0.036)	(0.034)	(0.036)	(0.034)	(0.036)	(0.034)
ΔExportToGDP	0.009	0.004	0.009	0.002	0.008	0.002
	(0.012)	(0.011)	(0.012)	(0.011)	(0.012)	(0.011)
ν		6.502		6.399		6.429
		(1.485)		(1.457)		(1.470)
logLik	-1234.141	-1199.966	-1233.930	-1199.251	-1233.088	-1198.969
AICc	2493.756	**2428.092**	2496.019	2429.421	2499.929	2434.604

Notes: This table shows estimation results for different model specifications (no spillovers, static spillovers, dynamic spillovers) based on normal or t-distribution as indicated in the column header. Standard errors are reported in brackets below coefficient estimates. Empty standard errors imply that the numerical matrix of standard errors is not positive and that the corresponding spillover model is not well suited to estimate this industrial sector. The Akaike information criterion (AICc) is depicted in bold for the model with the best fit (smallest value). The dependent variable is quarterly growth of the sector "Arts, entertainment, recreation and other services". The sample period spans 1996–2017. All variables are in quarterly growth rates. GDP and industrial sector growth rates are winsorized at the 1st and 99th percentile. See the data description in Table A.4.12 for more information on data sources.

Table A.4.4: Additional estimation results: construction

	no spillovers		static spillovers		dynamic spillovers	
	(1)	(2)	(3)	(4)	(5)	(6)
	normal	t	normal	t	normal	t
	ΔConstruction	ΔConstruction	ΔConstruction	ΔConstruction	ΔConstruction	ΔConstruction
ρ			0.213	0.228		
			(0.043)	(0.043)		
ω					0.051	0.039
						(0.071)
A					-0.000	0.008
					(0.004)	(0.012)
B					0.768	0.824
						(0.310)
$\ln(\sigma^2)$	-0.115	-0.292	-0.150	-0.324	-0.150	-0.327
	(0.048)	(0.074)	(0.048)	(0.072)	(0.048)	(0.072)
Constant	-0.100	-0.059	-0.085	-0.057	-0.086	-0.048
	(0.038)	(0.039)	(0.037)	(0.037)	(0.038)	(0.039)
Δ VIX	0.002	0.033	-0.040	-0.005	-0.040	0.000
	(0.120)	(0.119)	(0.118)	(0.115)	(0.118)	(0.115)
ΔEuroToDollar	1.507	1.408	1.462	1.339	1.457	1.376
	(0.656)	(0.653)	(0.644)	(0.619)	(0.640)	(0.622)
Δ Productivity	0.152	0.147	0.132	0.129	0.132	0.126
	(0.029)	(0.028)	(0.029)	(0.028)	(0.029)	(0.028)
ΔCreditToNonFinancialSector	-0.058	-0.051	-0.050	-0.045	-0.051	-0.043
	(0.017)	(0.017)	(0.017)	(0.016)	(0.017)	(0.017)
ΔConsumerConfidence	0.118	0.145	0.085	0.106	0.085	0.104
	(0.060)	(0.059)	(0.060)	(0.058)	(0.060)	(0.058)
ΔGFCF	0.021	0.016	0.019	0.015	0.019	0.015
	(0.004)	(0.004)	(0.004)	(0.004)	(0.004)	(0.004)
ΔLabourForce	0.172	0.149	0.153	0.136	0.153	0.134
	(0.060)	(0.060)	(0.059)	(0.059)	(0.059)	(0.058)
ΔGovernmentExpenditure	0.064	0.053	0.064	0.052	0.064	0.052
	(0.034)	(0.033)	(0.034)	(0.033)	(0.034)	(0.033)
ΔExportToGDP	0.001	-0.005	-0.007	-0.010	-0.007	-0.011
	(0.012)	(0.012)	(0.012)	(0.012)	(0.012)	(0.012)
ν		11.903		12.481		12.220
		(3.736)		(3.805)		(3.691)
logLik	-1198.092	-1186.048	-1186.254	-1173.074	-1186.253	-1172.761
AICc	2421.657	2400.257	2400.669	**2377.066**	2406.259	2382.189

Notes: This table shows estimation results for different model specifications (no spillovers, static spillovers, dynamic spillovers) based on normal or t-distribution as indicated in the column header. Standard errors are reported in brackets below coefficient estimates. Empty standard errors imply that the numerical matrix of standard errors is not positive and that the corresponding spillover model is not well suited to estimate this industrial sector. The Akaike information criterion (AICc) is depicted in bold for the model with the best fit (smallest value). The dependent variable is quarterly growth of the sector "Construction". The sample period spans 1996–2017. All variables are in quarterly growth rates. GDP and industrial sector growth rates are winsorized at the 1st and 99th percentile. See the data description in Table A.4.12 for more information on data sources.

Table A.4.5: Additional estimation results: financial and insurance activities

	no spillovers		static spillovers		dynamic spillovers	
	(1)	(2)	(3)	(4)	(5)	(6)
	normal	t	normal	t	normal	t
	ΔFinance & Insurance	ΔFinance & Insurance	ΔFinance & Insurance	ΔFinance & Insurance	ΔFinance & Insurance	ΔFinance & Insurance
ρ			0.161	0.174		
			(0.046)	(0.047)		
ω					0.004	0.307
					(0.000)	(0.110)
A					-0.012	0.014
						(0.021)
B					0.976	-0.712
					(0.002)	(0.363)
$\ln(\sigma^2)$	-0.048	-0.184	-0.066	-0.210	-0.075	-0.211
	(0.048)	(0.070)	(0.048)	(0.071)	(0.048)	(0.071)
Constant	-0.056	-0.050	-0.050	-0.044	-0.048	-0.040
	(0.039)	(0.039)	(0.039)	(0.038)	(0.035)	(0.039)
Δ VIX	-0.000	-0.038	0.010	-0.016	0.010	-0.019
	(0.124)	(0.123)	(0.123)	(0.120)	(0.120)	(0.120)
ΔEuroToDollar	0.165	0.128	0.154	0.103	0.152	0.144
	(0.678)	(0.670)	(0.672)	(0.655)	(0.669)	(0.653)
Δ Productivity	0.134	0.118	0.135	0.118	0.147	0.118
	(0.030)	(0.030)	(0.029)	(0.029)	(0.029)	(0.030)
ΔCreditToNonFinancialSector	0.011	0.008	0.008	0.003	-0.008	0.003
	(0.018)	(0.018)	(0.018)	(0.017)	(0.018)	(0.017)
ΔConsumerConfidence	0.099	0.077	0.100	0.080	0.109	0.080
	(0.062)	(0.062)	(0.062)	(0.061)	(0.062)	(0.061)
ΔGFCF	-0.003	-0.002	-0.003	-0.002	-0.003	-0.002
	(0.004)	(0.004)	(0.004)	(0.004)	(0.004)	(0.004)
ΔLabourForce	0.006	0.017	0.001	0.009	-0.011	0.006
	(0.062)	(0.059)	(0.062)	(0.058)	(0.061)	(0.058)
ΔGovernmentExpenditure	0.016	0.011	0.009	0.006	0.014	0.006
	(0.036)	(0.035)	(0.035)	(0.035)	(0.035)	(0.035)
ΔExportToGDP	0.001	0.004	0.001	0.004	-0.001	0.004
	(0.012)	(0.012)	(0.012)	(0.012)	(0.012)	(0.012)
ν		15.432		14.569		14.594
		(5.646)		(5.171)		(5.202)
logLik	-1227.501	-1219.476	-1221.485	-1213.063	-1220.851	-1212.825
AICc	2480.475	2467.112	2471.130	**2457.044**	2475.455	2462.316

Notes: This table shows estimation results for different model specifications (no spillovers, static spillovers, dynamic spillovers) based on normal or t-distribution as indicated in the column header. Standard errors are reported in brackets below coefficient estimates. Empty standard errors imply that the numerical matrix of standard errors is not positive and that the corresponding spillover model is not well suited to estimate this industrial sector. The Akaike information criterion (AICc) is depicted in bold for the model with the best fit (smallest value). The dependent variable is quarterly growth of the sector "Financial and insurance activities". The sample period spans 1996–2017. All variables are in quarterly growth rates. GDP and industrial sector growth rates are winsorized at the 1st and 99th percentile. See the data in Table A.4.12 description for more information on data sources.

Table A.4.6: Additional estimation results: industry (except construction)

	no spillovers		static spillovers		dynamic spillovers	
	(1)	(2)	(3)	(4)	(5)	(6)
	normal	t	normal	t	normal	t
	ΔIndustry	ΔIndustry	ΔIndustry	ΔIndustry	ΔIndustry	ΔIndustry
ρ			0.343	0.223		
			(0.035)	(0.044)		
ω					0.358	0.185
					(0.078)	(0.069)
A					0.025	0.031
					(0.012)	(0.018)
B					-0.621	-0.042
					(0.240)	(0.263)
$\ln(\sigma^2)$	-0.442	-0.897	-0.562	-0.896	-0.588	-0.886
	(0.048)	(0.086)	(0.048)	(0.082)	(0.050)	(0.081)
Constant	-0.129	-0.121	-0.103	-0.123	-0.083	-0.116
	(0.032)	(0.030)	(0.030)	(0.029)	(0.031)	(0.030)
Δ VIX	-0.151	-0.094	-0.073	-0.056	-0.099	-0.074
	(0.102)	(0.082)	(0.096)	(0.081)	(0.095)	(0.082)
ΔEuroToDollar	-1.505	-0.621	-0.970	-0.592	-0.759	-0.528
	(0.557)	(0.476)	(0.527)	(0.460)	(0.500)	(0.482)
Δ Productivity	0.385	0.451	0.334	0.428	0.324	0.424
	(0.024)	(0.031)	(0.024)	(0.030)	(0.024)	(0.030)
ΔCreditToNonFinancialSector	-0.093	-0.036	-0.074	-0.035	-0.068	-0.030
	(0.015)	(0.014)	(0.014)	(0.014)	(0.014)	(0.014)
ΔConsumerConfidence	0.154	0.108	0.040	0.059	0.038	0.048
	(0.051)	(0.044)	(0.050)	(0.044)	(0.050)	(0.045)
ΔGFCF	0.004	0.000	0.002	-0.000	0.000	-0.000
	(0.004)	(0.003)	(0.004)	(0.003)	(0.004)	(0.003)
ΔLabourForce	0.047	0.042	0.017	0.036	0.015	0.025
	(0.051)	(0.047)	(0.048)	(0.047)	(0.048)	(0.046)
ΔGovernmentExpenditure	0.025	0.041	0.024	0.034	0.038	0.041
	(0.029)	(0.026)	(0.028)	(0.025)	(0.027)	(0.026)
ΔExportToGDP	0.083	0.057	0.057	0.051	0.052	0.050
	(0.010)	(0.010)	(0.010)	(0.010)	(0.009)	(0.010)
ν		5.865		6.792		7.641
		(1.181)		(1.482)		(1.814)
logLik	-1054.129	-980.087	-1011.261	-968.525	-995.992	-960.739
AICc	2133.731	1988.335	2050.681	1967.969	2025.738	**1958.145**

Notes: This table shows estimation results for different model specifications (no spillovers, static spillovers, dynamic spillovers) based on normal or t-distribution as indicated in the column header. Standard errors are reported in brackets below coefficient estimates. Empty standard errors imply that the numerical matrix of standard errors is not positive and that the corresponding spillover model is not well suited to estimate this industrial sector. The Akaike information criterion (AICc) is depicted in bold for the model with the best fit (smallest value). The dependent variable is quarterly growth of the sector "Industry (except construction)". The sample period spans 1996–2017. All variables are in quarterly growth rates. GDP and industrial sector growth rates are winsorized at the 1st and 99th percentile. See the data description in Table A.4.12 for more information on data sources.

Table A.4.7: Additional estimation results: information and communication

	no spillovers		static spillovers		dynamic spillovers	
	(1)	(2)	(3)	(4)	(5)	(6)
	normal	t	normal	t	normal	t
	ΔInformation & Communication	ΔInformation & Communication	ΔInformation & Communication	ΔInformation & Communication	ΔInformation & Communication	ΔInformation & Communication
ρ			0.242	0.265		
			(0.043)	(0.044)		
ω					0.084	0.111
						(0.089)
A					-0.056	0.043
						(0.019)
B					0.671	0.543
					(0.009)	(0.327)
$\ln(\sigma^2)$	-0.078	-0.391	-0.122	-0.420	-0.136	-0.437
	(0.048)	(0.084)	(0.048)	(0.081)	(0.048)	(0.082)
Constant	-0.061	-0.129	-0.048	-0.090	0.011	-0.107
	(0.039)	(0.038)	(0.038)	(0.036)	(0.028)	(0.038)
Δ VIX	0.263	0.164	0.173	0.097	0.214	0.081
	(0.122)	(0.114)	(0.120)	(0.107)	(0.123)	(0.104)
ΔEuroToDollar	1.386	0.952	1.126	0.765	0.579	0.889
	(0.668)	(0.666)	(0.655)	(0.616)	(0.647)	(0.599)
Δ Productivity	0.193	0.200	0.186	0.190	0.175	0.187
	(0.029)	(0.032)	(0.029)	(0.030)	(0.028)	(0.030)
ΔCreditToNonFinancialSector	-0.018	-0.003	-0.020	-0.005	-0.013	-0.004
	(0.018)	(0.017)	(0.017)	(0.017)	(0.016)	(0.017)
ΔConsumerConfidence	-0.036	-0.077	-0.060	-0.097	-0.054	-0.101
	(0.061)	(0.056)	(0.060)	(0.055)	(0.059)	(0.055)
ΔGFCF	-0.003	-0.005	-0.004	-0.005	-0.004	-0.005
	(0.004)	(0.004)	(0.004)	(0.004)	(0.004)	(0.003)
ΔLabourForce	0.027	0.023	0.010	0.010	0.008	0.009
	(0.061)	(0.061)	(0.060)	(0.059)	(0.059)	(0.059)
ΔGovernmentExpenditure	-0.003	-0.003	-0.008	-0.010	0.001	-0.011
	(0.035)	(0.035)	(0.034)	(0.034)	(0.034)	(0.034)
ΔExportToGDP	0.023	0.017	0.014	0.009	0.011	0.006
	(0.012)	(0.012)	(0.012)	(0.011)	(0.012)	(0.011)
ν		7.010		7.630		7.257
		(1.635)		(1.793)		(1.655)
logLik	-1214.319	-1183.970	-1199.730	-1168.358	-1198.060	-1165.225
AICc	2454.112	2396.101	2427.619	2367.636	2429.873	**2367.117**

Notes: This table shows estimation results for different model specifications (no spillovers, static spillovers, dynamic spillovers) based on normal or t-distribution as indicated in the column header. Standard errors are reported in brackets below coefficient estimates. Empty standard errors imply that the numerical matrix of standard errors is not positive and that the corresponding spillover model is not well suited to estimate this industrial sector. The Akaike information criterion (AICc) is depicted in bold for the model with the best fit (smallest value). The dependent variable is quarterly growth of the sector "Information and communication". The sample period spans 1996–2017. All variables are in quarterly growth rates. GDP and industrial sector growth rates are winsorized at the 1st and 99th percentile. See the data description in Table A.4.12 for more information on data sources.

Table A.4.8: Additional estimation results: professional, scientific and tech activities

	no spillovers		static spillovers		dynamic spillovers	
	(1)	(2)	(3)	(4)	(5)	(6)
	normal	t	normal	t	normal	t
	ΔProfessio-nal, Scientific & Tech Activities	ΔProfessio-nal, Scientific & Tech Activities	ΔProfessio-nal, Scientific & Tech Activities	ΔProfessio-nal, Scientific & Tech Activities	ΔProfessio-nal, Scientific & Tech Activities	ΔProfessio-nal, Scientific & Tech Activities
ρ			0.336	0.302		
			(0.038)	(0.043)		
ω					0.106	0.121
					(0.055)	(0.061)
A					0.042	0.043
					(0.009)	(0.013)
B					0.600	0.546
					(0.154)	(0.173)
$\ln(\sigma^2)$	-0.113	-0.369	-0.213	-0.384	-0.230	-0.386
	(0.048)	(0.079)	(0.048)	(0.074)	(0.048)	(0.074)
Constant	-0.071	-0.001	-0.050	-0.023	-0.001	0.007
	(0.038)	(0.037)	(0.036)	(0.035)	(0.037)	(0.036)
Δ VIX	-0.010	-0.054	-0.033	-0.045	0.119	0.038
	(0.120)	(0.116)	(0.114)	(0.108)	(0.107)	(0.109)
ΔEuroToDollar	0.058	0.057	0.171	0.139	0.929	0.489
	(0.656)	(0.670)	(0.624)	(0.621)	(0.516)	(0.585)
Δ Productivity	0.153	0.116	0.125	0.111	0.107	0.103
	(0.029)	(0.027)	(0.028)	(0.027)	(0.027)	(0.026)
ΔCreditToNonFinancialSector	-0.074	-0.059	-0.061	-0.056	-0.049	-0.049
	(0.017)	(0.016)	(0.017)	(0.016)	(0.016)	(0.016)
ΔConsumerConfidence	0.114	0.115	0.080	0.075	0.083	0.074
	(0.060)	(0.059)	(0.058)	(0.057)	(0.057)	(0.056)
ΔGFCF	0.003	-0.001	0.001	-0.001	-0.001	-0.002
	(0.004)	(0.004)	(0.004)	(0.004)	(0.004)	(0.004)
ΔLabourForce	0.078	0.055	0.040	0.044	0.016	0.028
	(0.060)	(0.056)	(0.057)	(0.055)	(0.057)	(0.054)
ΔGovernmentExpenditure	0.040	0.049	0.042	0.049	0.050	0.050
	(0.034)	(0.033)	(0.033)	(0.032)	(0.033)	(0.032)
ΔExportToGDP	0.059	0.039	0.039	0.032	0.026	0.027
	(0.012)	(0.012)	(0.011)	(0.012)	(0.011)	(0.011)
ν		8.721		11.919		12.981
		(2.278)		(3.810)		(4.449)
logLik	-1198.953	-1174.457	-1164.015	-1153.135	-1155.719	-1147.117
AICc	2423.379	2377.073	2356.191	2337.188	2345.191	**2330.900**

Notes: This table shows estimation results for different model specifications (no spillovers, static spillovers, dynamic spillovers) based on normal or t-distribution as indicated in the column header. Standard errors are reported in brackets below coefficient estimates. Empty standard errors imply that the numerical matrix of standard errors is not positive and that the corresponding spillover model is not well suited to estimate this industrial sector. The Akaike information criterion (AICc) is depicted in bold for the model with the best fit (smallest value). The dependent variable is quarterly growth of the sector "Professional, scientific and tech activities". The sample period spans 1996–2017. All variables are in quarterly growth rates. GDP and industrial sector growth rates are winsorized at the 1st and 99th percentile. See the data description in Table A.4.12 for more information on data sources.

Table A.4.9: Additional estimation results: public administration, deference, education, human health and social work

	no spillovers		static spillovers		dynamic spillovers	
	(1)	(2)	(3)	(4)	(5)	(6)
	normal	t	normal	t	normal	t
	ΔPublic Administration	ΔPublic Administration	ΔPublic Administration	ΔPublic Administration	ΔPublic Administration	ΔPublic Administration
ρ			-0.011	-0.022		
			(0.052)	(0.052)		
ω					-0.009	-0.032
					(0.062)	(0.076)
A					0.021	0.020
					(0.026)	(0.029)
B					-0.073	-0.414
					(0.758)	
$\ln(\sigma^2)$	-0.082	-0.249	-0.082	-0.249	-0.083	-0.249
	(0.048)	(0.074)	(0.048)	(0.074)	(0.048)	(0.073)
Constant	-0.121	-0.146	-0.121	-0.146	-0.122	-0.150
	(0.039)	(0.038)	(0.039)	(0.038)	(0.039)	(0.038)
Δ VIX	-0.007	-0.079	-0.007	-0.081	-0.010	-0.083
	(0.122)	(0.115)	(0.122)	(0.115)	(0.121)	(0.114)
ΔEuroToDollar	-0.787	-0.880	-0.791	-0.890	-0.747	-0.770
	(0.666)	(0.618)	(0.667)	(0.619)	(0.672)	(0.630)
Δ Productivity	0.041	0.035	0.041	0.034	0.041	0.036
	(0.029)	(0.029)	(0.029)	(0.029)	(0.029)	(0.029)
ΔCreditToNonFinancialSector	-0.005	-0.001	-0.005	-0.001	-0.005	-0.000
	(0.018)	(0.018)	(0.018)	(0.018)	(0.018)	(0.018)
ΔConsumerConfidence	-0.014	-0.026	-0.014	-0.026	-0.012	-0.026
	(0.061)	(0.058)	(0.061)	(0.058)	(0.062)	(0.059)
ΔGFCF	-0.002	-0.001	-0.002	-0.001	-0.001	-0.001
	(0.004)	(0.004)	(0.004)	(0.004)	(0.004)	(0.004)
ΔLabourForce	0.067	0.087	0.066	0.087	0.068	0.093
	(0.061)	(0.063)	(0.0619	(0.063)	(0.061)	(0.063)
ΔGovernmentExpenditure	0.261	0.282	0.261	0.282	0.262	0.282
	(0.035)	(0.037)	(0.035)	(0.037)	(0.035)	(0.037)
ΔExportToGDP	-0.013	-0.009	-0.014	-0.010	-0.013	-0.009
	(0.012)	(0.012)	(0.012)	(0.012)	(0.012)	(0.012)
ν		12.532		12.496		12.416
		(4.112)		(4.095)		(4.012)
logLik	-1212.554	-1202.135	-1212.531	-1202.043	-1212.203	-1201.847
AICc	2450.581	**2432.430**	2453.222	2435.005	2458.160	2440.361

Notes: This table shows estimation results for different model specifications (no spillovers, static spillovers, dynamic spillovers) based on normal or t-distribution as indicated in the column header. Standard errors are reported in brackets below coefficient estimates. Empty standard errors imply that the numerical matrix of standard errors is not positive and that the corresponding spillover model is not well suited to estimate this industrial sector. The Akaike information criterion (AICc) is depicted in bold for the model with the best fit (smallest value). The dependent variable is quarterly growth of the sector "Public administration, deference, education, human health and social work". The sample period spans 1996–2017. All variables are in quarterly growth rates. GDP and industrial sector growth rates are winsorized at the 1st and 99th percentile. See the data description in Table A.4.12 for more information on data sources.

Table A.4.10: Additional estimation results: real estate activities

	no spillovers		static spillovers		dynamic spillovers	
	(1)	(2)	(3)	(4)	(5)	(6)
	normal	t	normal	t	normal	t
	ΔReal Estate	ΔReal Estate	ΔReal Estate	ΔReal Estate	ΔReal Estate	ΔReal Estate
ρ			0.024	0.035		
			(0.050)	(0.052)		
ω					0.008	0.019
					(0.009)	(0.040)
A					-0.019	0.022
					(0.014)	(0.023)
B					0.810	0.372
					(0.153)	(0.458)
$\ln(\sigma^2)$	-0.032	-0.187	-0.032	-0.188	-0.036	-0.193
	(0.048)	(0.073)	(0.048)	(0.073)	(0.048)	(0.073)
Constant	-0.041	-0.068	-0.040	-0.066	-0.025	-0.072
	(0.040)	(0.039)	(0.040)	(0.039)	(0.041)	(0.040)
Δ VIX	0.025	0.083	0.025	0.081	0.057	0.068
	(0.125)	(0.124)	(0.125)	(0.124)	(0.127)	(0.124)
ΔEuroToDollar	0.084	-0.098	0.072	-0.119	0.140	-0.222
	(0.683)	(0.656)	(0.684)	(0.654)	(0.692)	(0.658)
Δ Productivity	0.106	0.094	0.105	0.093	0.103	0.092
	(0.030)	(0.031)	(0.030)	(0.031)	(0.030)	(0.031)
ΔCreditToNonFinancialSector	-0.020	-0.015	-0.020	-0.015	-0.017	-0.016
	(0.018)	(0.018)	(0.018)	(0.018)	(0.018)	(0.018)
ΔConsumerConfidence	-0.141	-0.111	-0.140	-0.111	-0.145	-0.114
	(0.063)	(0.062)	(0.063)	(0.062)	(0.063)	(0.062)
ΔGFCF	-0.003	-0.002	-0.003	-0.002	-0.004	-0.002
	(0.005)	(0.004)	(0.005)	(0.004)	(0.005)	(0.004)
ΔLabourForce	0.061	0.041	0.062	0.042	0.056	0.040
	(0.063)	(0.061)	(0.063)	(0.061)	(0.063)	(0.061)
ΔGovernmentExpenditure	0.017	0.031	0.017	0.031	0.015	0.029
	(0.036)	(0.036)	(0.036)	(0.036)	(0.036)	(0.036)
ΔExportToGDP	0.011	0.017	0.011	0.017	0.011	0.016
	(0.012)	(0.012)	(0.012)	(0.012)	(0.012)	(0.012)
ν		13.456		13.446		13.021
		(4.644)		(4.620)		(4.375)
logLik	-1234.731	-1225.420	-1234.618	-1225.193	-1233.297	-1224.760
AICc	2494.935	**2479.001**	2497.395	2481.306	2500.348	2486.187

Notes: This table shows estimation results for different model specifications (no spillovers, static spillovers, dynamic spillovers) based on normal or t-distribution as indicated in the column header. Standard errors are reported in brackets below coefficient estimates. Empty standard errors imply that the numerical matrix of standard errors is not positive and that the corresponding spillover model is not well suited to estimate this industrial sector. The Akaike information criterion (AICc) is depicted in bold for the model with the best fit (smallest value). The dependent variable is quarterly growth of the sector "Real estate activities". The sample period spans 1996–2017. All variables are in quarterly growth rates. GDP and industrial sector growth rates are winsorized at the 1st and 99th percentile. See the data description in Table A.4.12 for more information on data sources.

Table A.4.11: Additional estimation results: wholesale and retail trade, transport, accommodation and food

	no spillovers		static spillovers		dynamic spillovers	
	(1)	(2)	(3)	(4)	(5)	(6)
	normal	t	normal	t	normal	t
	ΔWholesale & Retail Trade	ΔWholesale & Retail Trade	ΔWholesale & Retail Trade	ΔWholesale & Retail Trade	ΔWholesale & Retail Trade	ΔWholesale & Retail Trade
ρ			0.284	0.250		
			(0.038)	(0.044)		
ω					0.041	0.048
					(0.034)	(0.046)
A					0.031	0.036
					(0.008)	(0.017)
B					0.724	0.717
					(0.174)	(0.219)
$\ln(\sigma^2)$	-0.240	-0.564	-0.315	-0.573	-0.337	-0.568
	(0.048)	(0.083)	(0.048)	(0.080)	(0.048)	(0.078)
Constant	-0.122	-0.056	-0.104	-0.077	-0.056	-0.052
	(0.036)	(0.035)	(0.034)	(0.033)	(0.035)	(0.033)
Δ VIX	0.117	0.168	0.092	0.133	0.147	0.173
	(0.112)	(0.107)	(0.108)	(0.101)	(0.106)	(0.101)
ΔEuroToDollar	-0.923	0.145	-0.500	-0.047	0.095	0.229
	(0.616)	(0.591)	(0.596)	(0.558)	(0.594)	(0.555)
Δ Productivity	0.263	0.211	0.233	0.206	0.215	0.201
	(0.027)	(0.031)	(0.026)	(0.029)	(0.026)	(0.029)
ΔCreditToNonFinancialSector	-0.082	-0.053	-0.067	-0.053	-0.057	-0.050
	(0.016)	(0.016)	(0.016)	(0.016)	(0.016)	(0.016)
ΔConsumerConfidence	0.209	0.179	0.135	0.141	0.115	0.128
	(0.057)	(0.052)	(0.056)	(0.052)	(0.055)	(0.052)
ΔGFCF	0.008	0.002	0.006	0.003	0.004	0.002
	(0.004)	(0.004)	(0.004)	(0.003)	(0.004)	(0.003)
ΔLabourForce	0.186	0.137	0.154	0.128	0.152	0.135
	(0.056)	(0.052)	(0.055)	(0.051)	(0.054)	(0.051)
ΔGovernmentExpenditure	0.024	0.049	0.022	0.046	0.028	0.046
	(0.032)	(0.031)	(0.031)	(0.031)	(0.031)	(0.030)
ΔExportToGDP	0.068	0.053	0.054	0.050	0.040	0.042
	(0.011)	(0.011)	(0.011)	(0.011)	(0.011)	(0.011)
ν		7.054		8.315		9.201
		(1.630)		(2.123)		(2.494)
logLik	-1142.829	-1107.144	-1116.689	-1093.062	-1104.064	-1085.503
AICc	2311.132	2242.449	2261.539	2217.043	2241.881	**2207.672**

Notes: This table shows estimation results for different model specifications (no spillovers, static spillovers, dynamic spillovers) based on normal or t-distribution as indicated in the column header. Standard errors are reported in brackets below coefficient estimates. Empty standard errors imply that the numerical matrix of standard errors is not positive and that the corresponding spillover model is not well suited to estimate this industrial sector. The Akaike information criterion (AICc) is depicted in bold for the model with the best fit (smallest value). The dependent variable is quarterly growth of the sector "Wholesale and retail trade, transport, accommodation and food". The sample period spans 1996–2017. All variables are in quarterly growth rates. GDP and industrial sector growth rates are winsorized at the 1st and 99th percentile. See the data description in Table A.4.12 for more information on data sources.

Figure A.4.1: Time-varying spillover strength for all industrial sectors

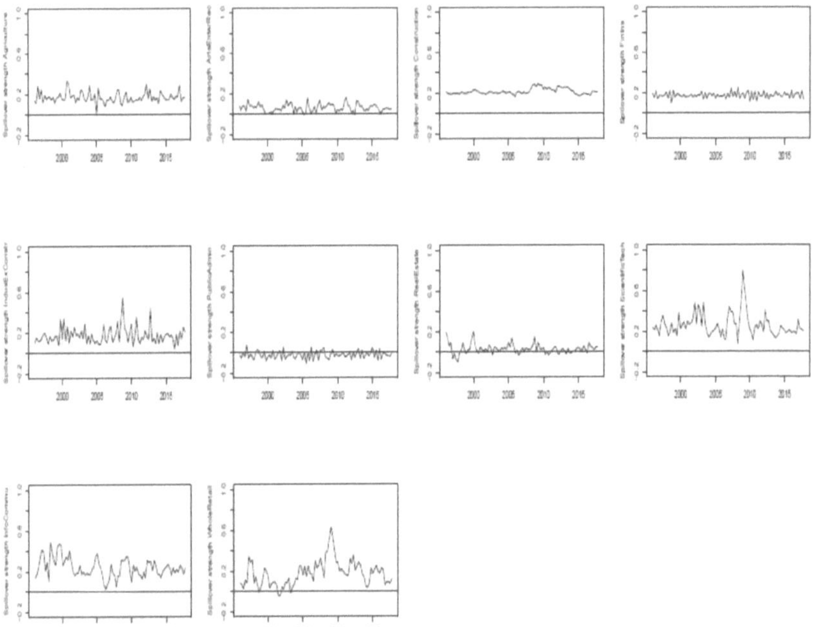

This graph shows the time-varying spillover strength for the model with industrial sector growth as the dependent variable and the sample period 1996-2017. The industrial sectors comprise in the first row: Agriculture, forestry and fishing. Arts, entertainment, recreation and other services. Construction. Financial and insurance activities. In the second row: Industry (except construction). Public administration, deference, education, human health and social work. Real estate activities. Professional, scientific and tech activities. In the third row: Information and communication. Wholesale and retail trade, transport, accommodation and food. The model is estimated based on score-driven model with Student's t-distributed disturbances.

Table A.4.12: Description and sources of variables

Variable	Description	Source
Dependent Variables (variables winsorized at 1st and 99th percentile)		
ΔGDP	Quarter-to-quarter growth rate of GDP in constant 2010 prices, seasonally adjusted.	OECD

ΔIndustry Sector	Quarter-to-quarter growth rates of gross value-added in constant 2010 prices, season and calendar adjusted. Industries according to Eurostat's A*10 industry breakdown are: Agriculture, forestry and fishing. Arts, entertainment, recreation and other services. Construction. Financial and insurance activities. Industry (except construction). Information and communication. Professional, scientific and tech activities. Public administration, defence, education, human health and social work. Real estate activities. Wholesale and retail trade, transport, accommodation and food.	Eurostat

Spatial Matrix

Spatial Matrix Weights	BIS locational banking statistics, total claims of all reporting banks (in all currencies and instruments) towards all sectors in counterparty country.	BIS

Control Variables

Variable	Description	Source
ΔProductivity	Labour productivity, growth rate, seasonally adjusted.	OECD
ΔConsumer Confidence	Consumer confidence indicator, growth rate, seasonally adjusted.	OECD
ΔGross Fixed Capital Formation (GFCF)	Gross fixed capital formation, growth rate, constant prices, seasonally adjusted.	OECD
ΔLabour Force	Labour force, growth rate, seasonally adjusted.	OECD
ΔGovernment Expenditure	Government final consumption expenditure, growth rate, constant prices, seasonally adjusted.	OECD
ΔCredit to Non-Financial Sector	Credit to private non-financial sector, growth rate, provided by all sectors, in percent of GDP.	BIS
ΔExports to GDP	Exports of goods and services, in percent of GDP.	Oxford Economics
ΔVIX	Volatility VIX, growth rate.	CBOE
ΔEuro-to-Dollar Exchange Rate	Euro to U.S. Dollar exchange rate, growth rate.	Thomson Reuters

References to Chapter 4

Abbott, A., Easaw, J. & Xing, T. (2008), 'Trade integration and business cycle convergence: Is the relation robust across time and space?', *The Scandinavian Journal of Economics* **110**(2), 403–417.

Acemoglu, D., Carvalho, V. M., Ozdaglar, A. & Tahbaz-Salehi, A. (2012), 'The network origins of aggregate fluctuations', *Econometrica* **80**(5), 1977–2016.

Arkolakis, C. & Ramanarayanan, A. (2009), 'Vertical specialization and international business cycle synchronization', *The Scandinavian Journal of Economics* **111**(4), 655–680.

Blasques, F., Koopman, S. J., Lucas, A. & Schaumburg, J. (2016), 'Spillover dynamics for systemic risk measurement using spatial financial time series models', *Journal of Econometrics* **195**(2), 211–223.

Braun, M. & Larrain, B. (2005), 'Finance and the business cycle: International, inter-industry evidence', *The Journal of Finance* **60**(3), 1097–1128.

Buch, C. M. (2000), 'Why do banks go abroad? Evidence from German data', *Financial Markets, Institutions & Instruments* **9**(1), 33–67.

Cerutti, E., Claessens, S. & Rose, A. K. (2019), 'How important is the global financial cycle? Evidence from capital flows', *IMF Economic Review* **67**(1), 24–60.

Cesa-Bianchi, A., Imbs, J. & Saleheen, J. (2019), 'Finance and synchronization', *Journal of International Economics* **116**, 74–87.

Chava, S. & Purnanandam, A. (2011), 'The effect of banking crisis on bank-dependent borrowers', *Journal of Financial Economics* **99**(1), 116–135.

Cravino, J. & Levchenko, A. A. (2017), 'Multinational firms and international business cycle transmission', *The Quarterly Journal of Economics* **132**(2), 921–962.

Creal, D., Koopman, S. J. & Lucas, A. (2011), 'A dynamic multivariate heavy-tailed model for time-varying volatilities and correlations', *Journal of Business and Economic Statistics* **29**(4), 552–563.

Creal, D., Koopman, S. J. & Lucas, A. (2013), 'Generalized autoregressive score models with applications', *Journal of Applied Econometrics* **28**, 777–795.

Denbee, E., Julliard, C., Li, Y. & Yuan, K. (2018), 'Network risk and key players: A structural analysis of interbank liquidity', *Fisher College of Business Working Paper 2018-03-011* .

Di Giovanni, J., Levchenko, A. A. & Mejean, I. (2018), 'The micro origins of international business-cycle comovement', *American Economic Review* **108**(1), 82–108.

Harvey, A. C. (2013), *Dynamic Models for Volatility and Heavy Tails*, Econometric Society Monographs, Cambridge University Press.

Herskovic, B., Kelly, B. T., Lustig, H. N. & Van Nieuwerburgh, S. (2018), 'Firm volatility in granular networks'. Chicago Booth Research Paper No. 12-56.

Hoffmann, M., Maslov, E., Sørensen, B. E. & Stewen, I. (2019), 'Channels of risk sharing in the Eurozone: What can banking and capital market union achieve?', *IMF Economic Review* **67**(3), 443–495.

Imbs, J. (2004), 'Trade, finance, specialization, and synchronization', *The Review of Economics and Statistics* **86**(3), 723–734.

Imbs, J. (2006), 'The real effects of financial integration', *Journal of International Economics* **68**(2), 296–324.

Kalemli-Ozcan, S., Papaioannou, E. & Perri, F. (2013), 'Global banks and crisis transmission', *Journal of International Economics* **89**(2), 495–510.

Kalemli-Ozcan, S., Papaioannou, E. & Peydró, J.-L. (2013), 'Financial regulation, financial globalization, and the synchronization of economic activity', *The Journal of Finance* **68**(3), 1179–1228.

Kleinert, J., Martin, J. & Toubal, F. (2015), 'The few leading the many: Foreign affiliates and business cycle comovement', *American Economic Journal: Macroeconomics* **7**(4), 134–59.

Kose, A. M., Prasad, E. S. & Terrones, M. E. (2003), 'How does globalization affect the synchronization of business cycles?', *American Economic Review* **93**(2), 57–62.

Kose, M. A., Otrok, C. & Prasad, E. (2012), 'Global business cycles: Convergence of decoupling?', *International Economic Review* **53**(2), 511–538.

Kroszner, R. S., Laeven, L. & Klingebiel, D. (2007), 'Banking crises, financial dependence, and growth', *Journal of Financial Economics* **84**(1), 187–228.

Laeven, L. & Valencia, F. (2013), 'The real effects of financial sector interventions during crises', *Journal of Money, Credit and Banking* **45**(1), 147–177.

Morgan, D. P., Rime, B. & Strahan, P. E. (2004), 'Bank integration and state business cycles', *The Quarterly Journal of Economics* **119**(4), 1555–1584.

Schnabel, I. & Seckinger, C. (2015), 'Financial fragmentation and economic growth in Europe'. CEPR Discussion Paper No. DP10805.

Tonzer, L. (2015), 'Cross-border interbank networks, banking risk and contagion', *Journal of Financial Stability* **18**, 19–32.

Chapter 5:

Physical Climate Change Risks and the Sovereign Creditworthiness of Emerging Economies

5.1 Introduction

As of 2020, human activities are estimated to have caused approximately 1.0°C of global warming compared to pre-industrial levels (IPCC 2018). Climate-related natural disasters, infectious diseases, species extinction and threats to economic prosperity as well as food, health and water supply are projected to increase dramatically with further warming. However, the IPCC (2018) also emphasizes that the 1.0°C increase witnessed so far has already led to more extreme weather events, changing natural systems and economic damages. Furthermore, the report states that the burden of climate change will be particularly heavy for developing countries in the global South.

In this paper, I exploit temperature fluctuations of past years which represent physical climate change risks in line with the 1.0°C warming witnessed so far. I contribute to the literature by linking these movements in temperature to the sovereign creditworthiness of, potentially climate-vulnerable, emerging market economies. Though the literature on the economic effects of temperature fluctuations is rich, the link to sovereign bond performances or sovereign risk has so far been missing.

Despite this gap in the literature, climate change can pose a significant threat for the creditworthiness of sovereigns according to several regulatory bodies. For instance, a report on the financial risks from climate change by the Bank of England (2018) states:

> *"The increasing frequency of severe weather events could also impact macroeconomic conditions through sustained damage to national infrastructure and weaken fundamental factors such as economic growth, employment, and inflation. This could have implications for the market price of sovereign debt for those countries most susceptible to the physical impacts of climate change."*[48]

Furthermore, rating agencies such as Moody's (2016) have started incorporating the credit implications of climate change for sovereign issuers.[49] These developments matter, as sovereign creditworthiness and associated bond costs are crucial for all governments. Rising borrowing costs compensate bondholders for higher risks, but can also push countries into crisis and default. Even in the absence of debt crises, any unit of currency that is spent on borrowing costs can no longer be used for other expenditures such as adaptions to climate change.

Therefore, I extend the literature on climate risks, in the form of temperature fluctuations, in connection with financial markets, in the form of sovereign bond returns. Figure 5.1 illustrates the main idea of my empirical approach. It depicts the mean annual temperature of the 54 countries in my panel from 1901 to 2018, showing an upward trend since the second half of the 20th century. The red line shows the constant temperature average from 1901 to 1950. From 1994 onward, which is the start of my estimation period and the shaded area in the graph, I calculate a country's temperature deviation from its 1901-1950 average. This temperature anomaly variable has a mean of $0.84°C$ which is close to the global warming trend of $1°C$ estimated by the IPCC (2018).

In my estimation, I follow the "new approach" outlined by Dell et al. (2014). Using monthly data for 54 emerging economies from 1994 to 2018, I regress market returns of the Emerging Market Bond Index (EMBI), a common measure for sovereign debt performance, on the described temperature anomaly fluctuations. I control for precipitation and include

---

[48]Similar remarks can be found by the ECB (2019), stating: *"sovereign risks could increase for countries with carbon-intensive industries."*

[49]Moreover, governments are increasingly facing legal consequences for not disclosing climate risks in their sovereign bond disclosures, as described in a Bloomberg (2020) article from 22 June 2020: *"Australia Sued For Not Disclosing Climate Risk in Sovereign Debt"*

Figure 5.1: Average annual temperature of 54 emerging economies in the sample from 1901-2018 and 1901-1950 temperature average

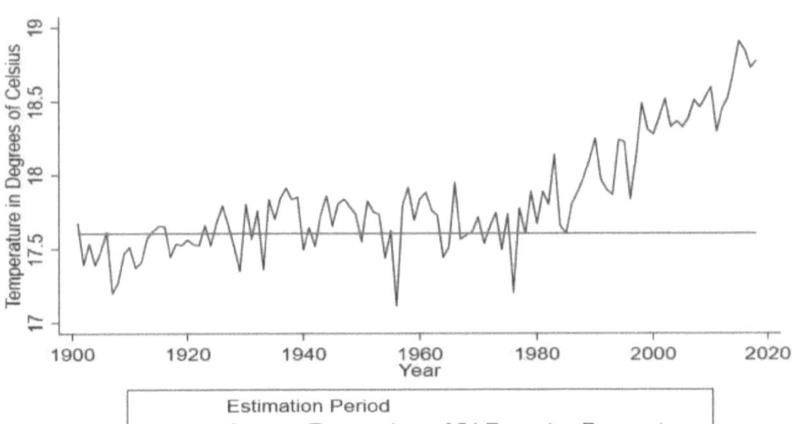

country and region-time fixed effects on the month-year level. The captured temperature shocks are thus idiosyncratic and account for weather trends common to each region. Building on a rich literature that links temperature increases to lower GDP growth in poorer and warmer countries (Burke et al. 2015, Dell et al. 2012), reduced firm productivity and output (Zhang et al. 2018, Adhvaryuy et al. 2020), decreasing labor supply (Graff Zivin & Neidell 2014) and more interpersonal and civil conflict (Hsiang et al. 2013), I empirically test the hypothesis if rising temperatures compared to a country's historical temperature average lead to lower sovereign debt performance (i.e. increasing sovereign risk).

My results indicate that the effect of rising temperature anomalies on sovereign credit-worthiness critically hinges on a country's economic and climatic profile: Warm countries are significantly more susceptible to temperature shocks than cold or mild-tempered countries, which is line with the results of Burke et al. (2015). For countries with very high average annual temperatures ($> 25°C$), a $1°C$ increase in monthly temperature compared to a country's historical average lowers EMBI returns by 0.464 percentage points on average. This effect corresponds to 11.9% of the EMBI returns' overall standard deviation. Thus, in a $2°C$ global warming scenario, EMBI returns (in percentage points) could be lowered for affected countries by roughly a quarter of their overall standard deviation. This magnitude is non-negligible and could lead to rising sovereign borrowing costs or even defaults for warmer countries in the next decades. Such out-of-sample projections must of course be treated carefully, as they abstain from countries' adaption strategies towards climate change but also from potentially non-linearly aggravating weather effects that are entailed by continuously rising temperatures (see Bolton et al. (2020)). However, if the past temperature anomaly shocks captured in this paper are any guidance, warm countries could bear a major burden from future temperature increases in the form of lower sovereign creditworthiness.

Related to the warmness of a country, I find that countries with lower temperature-seasonality suffer statistically and economically significantly more from temperature increases with respect to their sovereign risk level than countries with more volatile seasons. This result holds either for grouping countries into different bins of seasonality or for dividing the temperature anomaly measure by a country's standard deviation of monthly temperature.

Next, I exploit the monthly frequency of my data. I test if temperature shock effects differ in warmer compared to colder months or in summer compared to winter. However, after adjusting the months in southern hemisphere countries to the northern hemisphere scale, I find no statistical evidence that a temperature shock in warmer months has significantly different implications for sovereign creditworthiness than shocks in colder months, or differs in summer compared to winter. Put together with the previous evidence, this result suggests that the overall warm- or coldness of a country is what matters for temperature-induced sovereign risk, not the within-year seasonality of the weather.

Following the analysis of a country's climatic profile, I test if different economic sector specializations could be related to the strength of historical temperature shocks on sovereign debt performance. To this end, I interact the temperature anomaly measure with the specialization of a country in terms of agriculture, manufacturing, services or natural resources. However, these specifications do not yield any statistical patterns indicating that countries with higher agricultural shares on GDP, more service sector employees or larger rents from natural resources such as oil are more (or less) susceptible to temperature shocks with respect to their sovereign risk. My results do not rule out that potentially stranded industries, such as fossil fuels, may affect sovereign debt prices in the future, once their business models have become under stronger pressure. Still, the effect seems to be weak during my estimation period or not connected to temperature shocks.

What instead holds remarkably well throughout the analysis is the conditioning impact of institutional quality on temperature-induced sovereign risk. Countries with weaker rule of law, control of corruption, civil rights, democratic governments or less progressive tax systems face a statistically significantly stronger marginal effect of temperature increases that is detrimental to their sovereign creditworthiness. Next to these more traditional institutional variables, climate-related metrics yield a similar conclusion: Countries with lower values in the ND-Gain index, which measures both the adaptiveness and vulnerability of a country towards climate change, face significantly higher temperature shock effects on their sovereign risk level. Disentangling the ND-Gain index reveals that this effect is driven more by the adaptive readiness than the vulnerability part of the index. These results suggest that higher overall institutional quality, both traditional and climate-related, could improve the resilience and adaptiveness of emerging economies towards climate change.

I conduct encompassing robustness tests to demonstrate the stability of my results. These procedures include changing the fixed effects specification, dependent variable, historical average period and lag structure of temperature shocks. I also drop certain countries from the analysis, firstly if they have few EMBI data points, secondly if their landmass is among the ten largest countries. In addition, I test if more volatile weather periods can also impair sovereign bond performance, for which I find confirmation. Lastly, I analyze if the temperature effects changed after the Paris Agreement in December 2015, which does not seem to be the case.

In sum, my evidence suggests that historical temperature deviations, approximating physical climate change, lower sovereign bond performances (i.e. increase sovereign risk) significantly for countries that are: (i) warmer, (ii) have lower seasonality, (iii) and have lower institutional quality, both for traditional and climate-related metrics. I also find evidence that poorer countries suffer more from temperature shocks. However, these factors are correlated as poorer countries tend to have worse institutions. In addition, it is difficult to disentangle the long-run effects of climate zones on the creation of institutions or the wealth of nations (see Acemoglu et al. (2002) for a discussion).

I shed some light on these interrelations by combining all relevant channels, i.e. warmness, poverty and institutional quality, in one regression. My evidence suggests that the effect of poorer countries suffering stronger from temperature shocks is indeed driven by these countries' tendencies to have worse institutions. However, both the institutional and the warmness channel remain statistically significant in the same specification, suggesting that stronger institutions can provide resilience towards temperature shocks, independent of the warmness of a country.

Though any further disentanglement of these channels is beyond the scope of this paper, what matters for the policy implications is the finding that countries with warmer weather and lower institutional quality have so far been hit significantly harder by temperature anomaly shocks with respect to their sovereign creditworthiness. This result is an important extension to the still young literature on climate risks and financial markets. If past trends are any guidance, affected countries could face meaningful increases in their sovereign debt costs or even debt crises as climate change intensifies.

The rest of the paper is structured as follows. Section 5.2 provides a framework on how to think about physical climate change risk and its connection to sovereign risk. Section 5.3 introduces the data and provides summary statistics. In the following, Section 5.4 describes the empirical framework and main regression results. Section 5.5 investigates the climatic and economic profiles of countries and their connection to temperature-induced sovereign risk. The subsequent Section 5.6 provides encompassing robustness checks. Section 5.7 concludes.

5.2 Physical Climate Change Risk

5.2.1 Physical Climate Change Risk in Contrast to Transition Risk

The following section provides a framework on how to think about climate change risks in a sovereign bond context. Table 5.1 by the Bank of England (2018) depicts the distinction between *physical* and *transition* risks as the two main channels of how climate change can lead to economic impairments.

Physical risks describe the materializing damages from climate change. They can arise from extreme weather events or natural disasters such as droughts, wildfires, sea level rises or floods. Regions hit by such disasters can face losses in terms of human lives, critical infrastructure, food supply, firm assets or their capital stock (see also Bolton et al. (2020)). As further global warming likely entails irreversible tipping points, these damages could lead to

Table 5.1: Distinction between physical and transition climate change risks. Source: Bank of England (2018)

Risk Type	Implications for Credit	Implications for Markets	Implications for Business
Physical	Increasing flood risk to mortgage portfolios; declining agricultural output; increasing default rates	Severe weather events can lead to re-pricing of sovereign debt	Severe weather events can impact business continuity
Transition	Tightening efficiency standards impact property exposures; stranded assets impair loan portfolios; disruptive technology leads to auto finance losses	Tightening climate-related policy leads to re-pricing of securities and derivatives	Changing sentiment on climate issues leads to reputational risks

non-transitory, lasting disruptions (Ripple et al. 2019). According to the insurance data used by NGO Germanwatch (2019), the damages from extreme weather events worldwide between 1999 and 2018 amounted to \$3.54 trillion (in purchasing power parities). Physical climate risks can materialize as a mortgage risk for homeowners that lose their property, a credit risk for banks that lend to e.g. flood-impaired firms (Koetter et al. (2020)), an underwriting risk for insurance companies (Financial Stability Institute 2019) and, as demonstrated in this paper, a market risk for sovereigns bonds of countries most susceptible to the physical impacts of climate change.

In contrast, transition risks describe the adjustment towards a low-carbon economy and the expected damages and costs associated therewith. Therefore, these risks are more forward-looking as (expected) changes in environmental policies or sentiments could threaten, for instance, the business model of certain firms. Should investors reassess the viability of e.g. a fossil-energy-intensive industry as tougher climate laws are implemented, the stock price of affected firms might fall. Such a shock would likely spill-over to banks, pension funds and other investors with exposures towards stranded industries, which is referred to as a "carbon bubble" (see ESRB (2016) for an associated systemic risk analysis and Delis et al. (2018) for how banks price carbon bubble risks).

An example of transition risks in a government bond context that contrasts the physical risks in this paper is by Painter (2020). He shows that US municipalities that face stronger sea level increases in the future have higher issuance costs for their municipality bonds today. Because of its forward-looking nature, this effect demonstrates a transition risk. As projected climate damages from sea level increases rise over time, the results are driven by long-term bonds. In addition, the pricing effect increased around the release of the Stern report on climate change in 2006. Though not shown by Painter (2020), it could likely be the case that such re-pricing of climate-sensitive assets was even more pronounced in recent years as global warming became a major concern for the financial industry (see Boston Common Asset Management (2018) for a survey of global banks and Bolton & Kacperczyk (2020) for asset pricing effects of firms' CO2 emissions).

In contrast to forward-looking transition risks, this paper, and the literature on temperature effects in general, analyze already materialized impacts of past temperature fluctuations. Temperature increases are associated with extreme weather events or hotter years and influence economic activities along several dimensions, as the next section demonstrates. Of course, both risk channels cannot be isolated completely from another: A wildfire might entail vast economic damages (physical risk), but also change perceptions of investors regarding the susceptibility of the affected region towards more wildfires in the future (transition risk). It is beyond the scope of this paper to disentangle these risk effects. Nevertheless, I will label temperature fluctuations as a form of physical risk in the following due to their primary impact on current economic activities.

5.2.2 Physical Climate Change and Sovereign Creditworthiness

Temperature fluctuations have economic effects that can likely spill-over to sovereign risk. Dell et al. (2012) show that higher temperatures reduce the GDP growth rate of poorer countries. This effect is driven by lower agricultural and industrial value-added and increasing political instability during warmer years. Related, Burke et al. (2015) show that temperature has a non-linear effect on GDP growth, with warmer countries' economies being hit significantly more negatively by higher temperatures than colder or milder-tempered countries for which temperature increases are negligible or even beneficial. Heal & Park (2014) and Deryugina & Hsiang (2014) obtain similar results. Regarding this paper's research agenda, it is likely that macroeconomic fundamentals like GDP growth or related fiscal conditions impact sovereign bond pricing (see Hilscher & Nosbusch (2010) or Gupta et al. (2008)).

With respect to the microeconomic channels behind the temperature-GDP connection, Zhang et al. (2018) find that more hot days per year in a Chinese region significantly reduce output and productivity of local firms. Using climate prediction models, the authors derive that these effects could lower Chinese manufacturing output by 12% annually by 2050. Adhvaryuy et al. (2020), Cachon et al. (2012) and Somanathan et al. (2018) obtain similar evidence, confirming that labor becomes less productive with hotter days. In addition, Graff Zivin & Neidell (2014) demonstrate that individual labor supply decreases with more warm days in a year. Pankratz & Schiller (2019) show that climate shocks can negatively impact global production networks. One notable exception to this micro evidence is by Addoum et al. (2020) who find weak effects of temperature shocks on US firm sales.

Climate and weather patterns also influence conflict and political stability. Hsiang et al. (2013) summarize in a meta-study several contributions that link increasing temperatures to more interpersonal conflict and crime, but also riots, civil conflict or ultimately civil war (see also Burke et al. (2009)). Sovereign bond yields are known to respond to political conditions (Eichler 2014) and it is highly plausible for temperature-induced political instability to increase sovereign risk.

Though not every natural disaster can be directly linked to climate change, the IPCC (2018) projects climate-related disasters to increase with further global warming. Figure 5.2 depicts the total number of climate-related natural disasters such as floods, droughts and wildfires of the countries in my panel next to the average sample temperature from 1901 to 2018. There is a positive correlation between the rising occurrence of natural disasters and increasing temperature, however, this relationship is at least partially driven by better detection and recording of disasters. Nevertheless, the temperature anomaly measure in this paper picks up natural disasters to some extent, as shown in the next section, and it is intuitive to assume that severe disasters are detrimental to the economy and sovereign creditworthiness of a country (Felbermayr & Gröschl 2014).

As I use the market return of a financial asset as my dependent variable, it is worth noting that Bansal et al. (2016) demonstrate that most US equities have a negative exposure coefficient towards long-run temperature fluctuations. Temperature patterns and other climate-related measures are thus priced in financial assets (Bolton & Kacperczyk 2020).

The literature on the effects of temperature anomalies on sovereign creditworthiness is so far scarce, which is why this paper adds significant value to this debate. Next to cited work by Painter (2020), another paper that looks at the relationship between sovereign borrowing costs

Figure 5.2: Number of climate-related natural disasters and average temperature of panel countries

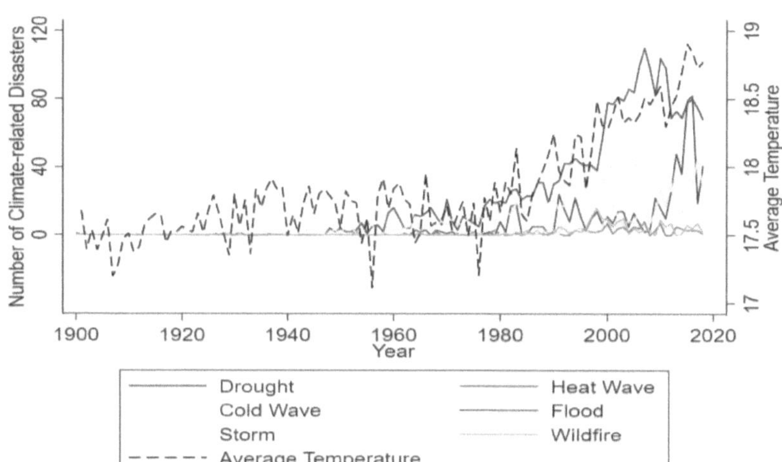

and climate change more general is by Kling et al. (2018). The authors regress bond costs on climate-related vulnerability metrics of countries, finding that more vulnerable countries pay higher debt costs. Though the specifics of the estimation strategy and the included countries differ, the results in my paper point in a similar direction.

5.3 Data and Descriptive Statistics

5.3.1 Sovereign Creditworthiness

I measure sovereign creditworthiness using the Emerging Market Bond Index Global (EMBI) provided by J.P. Morgan. EMBI data has several advantages: Included sovereign bonds are US Dollar-denominated which rules out exchange rate risk. Eligible debt must furthermore have more than one year to maturity and exceed an outstanding face value of $500 million. These features make EMBI data well standardized, liquid and widely-used to track sovereign debt performances of emerging economies.

The start of the EMBI Global at the beginning of 1994 determines my estimation period, which runs from 1994m1 to 2018m12. I collect monthly EMBI Global data for all countries available and calculate month-to-month returns using natural log differences. Positive returns imply improving sovereign creditworthiness.[50] I winsorize the returns at the 1st and 99th percentile to control for outliers. The panel is unbalanced because some countries enter only in later years. As some countries' EMBI series turn temporarily illiquid and hence constant in the index level, I drop all observations with a zero percent EMBI return. To make sure every country in the sample has sufficient variation, I only include those countries with liquid EMBI returns of at least six years (72 months). This criterion is not critical for my results, as shown in a robustness test. The final panel consists of 54 countries and can be found, together with region classifications from Dell et al. (2012), in Table 5.2. Definition and sources of all variables are in Table A.5.5.

Table 5.2: List of included countries and region classification

Region	Countries
Asia-Pacific	China, India, Indonesia, Malaysia, Mongolia, Pakistan, Philippines, Vietnam
Eastern Europe & Central Asia	Azerbaijan, Belarus, Croatia, Georgia, Hungary, Kazakhstan, Latvia, Lithuania, Poland, Romania, Russia, Serbia, Ukraine
Latin America & Caribbean	Argentina, Belize, Bolivia, Brazil, Chile, Colombia, Costa Rica, Dominican Republic, Ecuador, El Salvador, Guatemala, Jamaica, Mexico, Panama, Peru, Uruguay, Venezuela
Middle East & North Africa	Egypt, Iraq, Jordan, Lebanon, Morocco, Tunisia, Turkey
Sub-Sahara Africa	Angola, Gabon, Ghana, Ivory Coast, Namibia, Nigeria, Senegal, South Africa, Zambia

5.3.2 Temperature Data

I obtain average monthly temperature data for every panel country since 1901 from the Climate Research Unit (CRU). The data is land-weighted and based on an extensive network of interpolated weather station data (see Harris et al. (2020) for details).[51]

My main variable of interest, as graphically depicted in Figure 5.1, measures the difference in the observed temperature of a country during 1994m1-2018m12 towards this country's 1901-1950 historical temperature average of that month:

$$HistoricalTempAnomaly_{it} = Temperature_{it} - TempAverage_{i,t(1901-1950)} \qquad (17)$$

---

[50]I obtain somewhat stronger results using direct EMBI returns. However, the results also hold when using EMBI spread data as shown in the robustness section. Since both measures are market returns, their interpretation, except for the switched signs, is very similar.

[51]Data is freely available at: `https://crudata.uea.ac.uk/cru/data/hrg/cru_ts_4.03/`.

For instance, temperature in March of 2003 (year-month t) in Argentina (country i) is compared to the temperature of all months March of Argentina from 1901-1950. This historical temperature anomaly is a proxy for the degree of global warming witnessed so far. Table 5.3 listing the summary statistics shows a corresponding mean of $0.842°C$ for the full sample period. This value approaches the $1°C$ temperature increase estimated by the IPCC (2018) and lies well within their reported confidence range of $0.8°C$ to $1.2°C$. For the sample of temperature anomalies used in the main regressions, the mean is even at $0.896°C$, likely because several countries enter the estimation only in later and thus warmer years.

In line with the IPCC (2018)'s assessment that the $1°C$ warming so far has already led to impacts on natural and human systems and considering the evidence on the economic effects of temperature fluctuations gathered in Section 5.2, I interpret $HistoricalTempAnomaly_{it}$ as a measure for warmer than normal periods and extreme weather events. Some statistical confirmation for this perception comes from Table 5.4, showing that the mean of temperature anomalies is higher during periods of heat-related natural disasters such as droughts (0.898), droughts for which there is a damage estimate (0.946), wildfires (0.989) and heat waves (0.988).[52] In addition, I collect GDP growth, stock market and government primary surplus data, which is for the latter two variables only available for a subsample of countries. Table 5.5 shows that the overall mean of these economic conditions is differentiated during high and low temperature anomalies. The mean of stock returns (overall: 0.233) is lower if temperature increases are above the 75th percentile (-0.0478) and higher when temperature is below the 25th percentile (0.434). Similarly, primary surpluses (overall: -0.235) decrease during higher (-0.693) and rise during lower (0.400) temperature anomalies. Historical Temperature increases are thus responsive to both climate- and economy-related news.

I include an additional temperature variable in the main regressions:

$$DeviationAdjustedTempAnomaly_{it} = \frac{HistoricalTempAnomaly_{it}}{StandardDeviation(Temperature_{i,t(1901-1950)})}$$

$$(18)$$

I divide the anomaly measure by a country's historical standard deviation of monthly temperature. This adjustment is suggested by Dell et al. (2014) and applied, among others, by Barrios et al. (2010). It sets the temperature shock in relation to the usual variation in

---

[52]Wildfires or heat waves with reported damages also have higher averages but lower number of observation.

warm- or coldness of a country. In this way, temperature anomalies in countries with lower seasonality are stronger emphasized.

Table 5.3: Summary statistics of all variables

Variable	Obs.	Mean	Median	Std. Dev.	Min	Max
ΔEMBI	10,006	0.686	0.729	3.921	-16.23	13.47
ΔEMBI (baseline)	9,957	0.691	0.729	3.898	-16.23	13.47
HistoricalTempAnomaly	16,200	0.842	0.694	1.190	-5.514	8.830
HistoricalTempAnomaly (baseline)	9,957	0.896	0.742	1.129	-5.254	8.830
DeviationAdjustedTempAnomaly	16,200	0.355	0.223	0.514	-1.627	4.007
DeviationAdjustedTempAnomaly (baseline)	9,957	0.418	0.257	0.574	-1.627	4.007
Precipitation	16,200	0.0960	0.0610	0.0971	0	1.072
ΔVIX	16,146	0.0656	-0.0800	4.219	-10.85	15.35
ΔGlobalGovernmentBondIndex	16,146	0.367	0.307	1.817	-4.967	5.365
ΔUS-TermSpread	16,146	-0.00640	-0.0430	0.272	-0.551	0.800
ΔUS-CorporateRiskPremium	16,146	0.00347	-0.0257	0.480	-1.207	1.944
ΔUS-10-YearTreasuryYield	16,146	-0.00858	-0.0151	0.252	-0.744	0.635
AgricultureToGDP	16,032	10.36	8.300	6.952	2	38.96
ManufacturingToGDP	15,624	14.85	15.09	5.688	0.650	35.01
ServicesToGDP	16,032	51.42	52.65	9.533	10.57	75.85
ResourceRentsToGDP	15,348	7.069	2.950	9.825	0	64.15
RuleOfLaw	14,904	41.50	42.09	20.29	0	89.47
ControlOfCorruption	14,904	41.88	42.55	21.25	0.510	91.88
CivilRights	16,032	3.513	3	1.474	1	7
PoliticalRights	16,044	3.426	3	1.933	1	7
IncomeRedistribution	14,136	10.86	5.977	12.27	-21.01	46.68
Polity2	15,504	4.429	7	5.752	-9	10
DemocraticGovernments	15,264	5.971	7	3.469	0	10
AuthoritarianGovernments	15,264	1.481	0	2.459	0	9
ND-GAIN	15,552	47.72	47.02	6.398	32.61	62.80
ReadinessIndex	15,552	0.388	0.382	0.0851	0.181	0.609
VulnerabilityIndex	15,552	0.433	0.423	0.0541	0.322	0.569
GDPPerCapita	16,164	5,556	4,548	3,882	541.6	17,709
ΔEMBISpread	9,694	-0.589	-2.406	80.76	-310.2	371.9
ΔCDSSpread	4,328	1.911	-0.260	71.05	-281.1	410.7
HistoricalDeviationAnomaly	16,200	0.118	0.0508	0.452	-2.312	2.630
DebtToGDP	14,076	51.25	43.29	43.63	3.22	925.24
YearsSinceLastSovDebtRestructuring	16,200	22.71	23	13.18	0	36
PrimaryNetLendingGDP	11,340	-0.235	-0.431	3.72	-31.3	22.02
AgricultureLandShare	14,760	44.73	45.52	20.36	3.190	85.49
AgricultureEmploymentShare	16,200	26.93	23.23	15.73	0.0600	72.26
IndustrialEmploymentShare	16,200	21.09	21.40	6.441	6.240	35.99
ServicesEmploymentShare	16,200	51.97	52.57	12.48	20.55	78.80

Sample period is 1994:m1–2018:m12. Variables with Δ are in monthly growth rates and winsorized at the 1st and 99th percentile, all other variables are in levels. See Table A.5.5 for information on data sources.

Table 5.4: Mean of temperature anomaly measure during natural disaster periods

	Full Sample Mean	Drought	Droughts with reported Damage	Wildfire	Heat Wave	Cold Wave or severe Winter
Historical	0.842	0.898	0.946	0.989	0.988	0.0623
TempAnomaly	(16,200)	(716)	(505)	(132)	(82)	(269)

Mean of HistoricalTempAnomaly is shown for periods with occurring natural disasters in panel countries. Number of observations for each event is shown below in brackets.

Table 5.5: Mean of economic variables during high and low temperature anomalies

Economic Variable	Overall Mean	Mean if Historical Temperature Anomaly > 75th percentile	Mean if Historical Temperature Anomaly < 25th percentile
Stock Market Returns (38 Countries)	0.233	-0.0872	0.412
Quarterly GDP Growth (full sample)	0.981	0.937	1.061
Government Primary Surplus in % of GDP (42 Countries)	-0.235	-0.693	0.400

Mean of an economic variable is shown over all periods, for periods with high temperature anomalies and for periods with low temperature anomalies. HistoricalTempAnomaly is on monthly frequency for stock returns (75th percentile 1.382°C, 25th percentile 0.286°C), collapsed to quarterly frequency for GDP growth (75th percentile 1.245°C, 25th percentile 0.321°C), and collapsed to yearly frequency for government surplus (75th percentile 1.178°C, 25th percentile 0.447°C) to match the respective frequency.

5.4 Empirical Specification and Results

Following what Dell et al. (2014) call the "new approach", I estimate an OLS panel regression:

$$\Delta SovereignCreditworthiness_{it} = \beta TemperatureAnomaly_{it} + \delta Precip_{it} + \gamma_i + \gamma_{rt} + \epsilon_{it} \quad (19)$$

Natural log changes in the EMBI ($\Delta SovereignCreditworthiness_{it}$) are regressed on a temperature anomaly measure and fixed effects. The sample runs from 1994m1 to 2018m12 and consists of 54 countries. Temperature anomalies are either the difference of temperature from its historical average ($HistoricalTempAnomaly_{it}$) or the historical anomaly divided by monthly temperature standard deviation ($DeviationAdjustedTempAnomaly_{it}$) as described in Section 5.3.2. The size and statistical significance of β tests the hypothesis if and by how much temperature anomalies affect sovereign risk. Based on the gathered evidence in Section 5.2.2, I expect higher temperature anomalies leading to lower sovereign creditworthiness.

I include country fixed effects γ_i to control for time-invariant characteristics such as geography or culture. In addition, year-month fixed effects enter the regression and are interacted with the region classification of a country (γ_{rt}). This approach, suggested among others by Dell et al. (2014), makes sure that common trends, such as shared weather patterns in each region, are controlled for. It ensures that captured temperature shocks are idiosyncratic and local in nature. I apply different fixed effects in the robustness section and find stable results.

Importantly, I do not include any control variables on the country level such as stock returns or exchange rates. This decision is due to the explicit stance of the temperature literature against including any control variable that might be endogenous towards weather and climate variation (Dell et al. (2014), Burke et al. (2015)).[53] Given that stock returns are subject to similar temperature-productivity effects described in Section 5.2 and also unavailable on a liquid frequency for all panel countries, I abstain from including them. Following leading papers like Dell et al. (2012) and Burke et al. (2015), I only control for precipitation ($Precip_{it}$, also obtained from CRU) and include time-region fixed effects on the highest possible frequency (year-month). Standard errors are clustered on the country level.

In the context of this book, in particular the instrumental variable approach of Chapter 2, the choice to omit certain control variables could seem contradictory. In Chapter 2, I applied an IV approach and included a multitude of controls to rule out concerns on omitted variables between sovereign and bank risk. Chapter 5 consciously avoids additional controls. However, one has to keep in mind that the literature on temperature and the literature on the sovereign-bank loop are distinct in their empirical approaches. Leading papers of the temperature literature refer to the "bad control" chapter by Angrist & Pischke (2008), who warn against including control variables, that could themselves be an outcome of the variable of interest. Though this is a concern one could have for many empirical regressions, papers like Dell et al. (2014) and Burke et al. (2015) argue that the biasing effect of a bad control is particularly pressing in a temperature estimation. In the present case, stock returns are driven by the economic effects of temperature variation. They could also serve as an outcome variable instead of sovereign risk. Though the issue if a control is "bad" or not can be raised for many papers, in this book, I follow the leading articles of the respective literature and therefore exclude all biasing controls in the equations of this chapter.

Table 5.6 presents results from several versions of equation (19), including the baseline model. Column (1) introduces the historical temperature anomaly measure and both country and region-time fixed effects, but only on a yearly level. The temperature measure enters negative and statistically significant, but the overall explanatory power of the estimation is quite low. In column (2) I include precipitation on the country level and several international control variables such as changes in the VIX, the US term spread, US corporate risk spread, the 10-year US treasury yield and the returns of a general government bond index. The temperature anomaly coefficient remains negative and statistically significant to this addition.

---

[53]In their review article on climate and crime, Hsiang et al. (2013) explicitly exclude studies that use a potentially biasing control variable. See also the chapter "bad control" in Angrist & Pischke (2008).

Finally, I estimate the baseline model (column (3)) in which I introduce region times year-month fixed effects, which subsume all non-country specific controls. The explanatory power is now substantially larger, but the temperature anomaly measure is no longer statistically significant. This result might not be surprising, as several papers in the literature have statistically insignificant base effects[54] and one could hypothesize that only particularly susceptible countries are affected by temperature increases. Precipitation is statistically insignificant in the baseline and all following regressions.

The hypothesis that only susceptible countries respond to temperature shocks receives confirmation in columns (4) to (6). In these estimations, I repeat the specifications of columns (1)-(3) but replace the historical temperature anomaly with the deviation-adjusted temperature measure ($DeviationAdjustedTempAnomaly_{i,t}$). As described, this version emphasizes temperature shocks in countries with low seasonality. It enters negative and with a stable and strongly statistically significant coefficient (1% level) in all specifications. This result implies that rising temperature leads to a statistically significant decrease of sovereign creditworthiness for countries with low seasonality. Regarding the economic size, an increase of deviation-adjusted temperature anomalies by one standard deviation ($0.574°C$) leads to a 0.135%-point drop in EMBI returns. This magnitude corresponds to 3.47% of the standard deviation of EMBI returns in the estimation sample. While this effect is modest for now, the next section will investigate the susceptibility of countries towards temperature shocks in greater detail and identify more substantial effects.

5.5 Channels of Temperature-Sovereign Risk Connection

The previous literature established that temperature shocks can be particularly harmful for warmer or poorer countries or affect certain economic sectors like agriculture or industrial production (Burke et al. 2015, Dell et al. 2012). I investigate such channels with respect to their impact on sovereign risk. Specifically, I analyze the general warmness of a country (5.5.1), its seasonality (5.5.2), its within year weather fluctuations (5.5.3), its specialization towards different economic sectors (5.5.4), the effect of institutions (5.5.5) and ultimately a combination of all relevant channels (5.5.6) regarding their temperature-induced sovereign risk impact.

---

[54]For instance, Dell et al. (2012) also obtain a statistically insignificant baseline effect.

Table 5.6: Baseline results: temperature anomalies and sovereign risk

	(1) ΔEMBI	(2) ΔEMBI	(3) ΔEMBI	(4) ΔEMBI	(5) ΔEMBI	(6) ΔEMBI
HistoricalTempAnomaly	-0.0614**	-0.0798***	-0.0118			
	(0.0267)	(0.0269)	(0.0507)			
DeviationAdjustedTempAnomaly				-0.219***	-0.233***	-0.235***
				(0.0681)	(0.0633)	(0.0798)
Precipitation		0.197	0.223		0.110	0.0385
		(0.416)	(0.363)		(0.424)	(0.375)
ΔVIX		-0.151***			-0.151***	
		(0.0164)			(0.0164)	
ΔGlobalGovernmentBondIndex		0.0691*			0.0690*	
		(0.0370)			(0.0370)	
ΔUS-TermSpread		-0.102			-0.106	
		(0.199)			(0.199)	
ΔUS-CorporateRiskPremium		-2.445***			-2.443***	
		(0.186)			(0.186)	
ΔUS-10-YearTreasuryYield		-3.689***			-3.684***	
		(0.437)			(0.438)	
Constant	0.740***	0.688***	0.679***	0.777***	0.722***	0.786***
	(0.0239)	(0.0539)	(0.0631)	(0.0284)	(0.0560)	(0.0552)
Observations	10,006	10,006	9,957	10,006	10,006	9,957
R-squared	0.068	0.217	0.524	0.068	0.218	0.524
Country FE	Yes	Yes	Yes	Yes	Yes	Yes
Region$\times$Year FE	Yes	Yes	No	Yes	Yes	No
Region$\times$MonthYear FE	No	No	Yes	No	No	Yes

This table shows OLS estimation results of a panel of 54 countries from 1994m1 to 2018m12. ΔEMBI are monthly natural log returns of a country's EMBI index. HistoricalTempAnomaly is the difference between monthly temperature of a country and its 1901-1950 temperature average of the same month. Deviation-AdjustedTempAnomaly is the anomaly measure divided by a country's 1901-1950 average of temperature standard deviation. Precipitation is the country-specific average in 1000 mm units. ΔVIX, ΔUS-TermSpread (10-year treasury yield minus 3-month T-Bill yield), ΔUS-CorporateRiskPremium (high corporate bond yield minus investment grade corporate bond yield) and ΔUS-10-YearTreasuryYield are in simple first differences, ΔGlobalGovernmentBondIndex is in natural log differences. Standard errors (in parentheses) are clustered at the country level, ***, ** and * indicate statistical significance at the 1%, 5% and 10% level, respectively. See Table A.5.5 for variable definitions and sources.

Methodically, I either analyze these channels in an interaction model as follows:

$$\Delta SovereignCreditworthiness_{it} = \lambda_1 TemperatureAnomaly_{it} * Channel_{it} + \lambda_2 Channel_{it}$$
$$+ \beta TemperatureAnomaly_{it} + \delta Precip_{it} + \gamma_i + \gamma_{rt} + \epsilon_{it}$$
$$(20)$$

That is, the baseline estimation is repeated while $TemperatureAnomaly$ is interacted with the channel of interest, for instance institutional quality. I expect channels that increase the detrimental impact of temperature shocks on sovereign creditworthiness to enter with a negative, while factors that cushion the effect of temperature on sovereign bond performance to carry a positive coefficient sign.

Some of the analyzed channels could be endogenous towards temperature, such as the share of agriculture on the economy. However, as shown by Nizalova & Murtazashvili (2016) and Bun & Harrison (2019) even if one of such channels could be endogenous in the single term, the interacted effect with temperature anomalies can still yield a consistent estimate. This inference holds as long as one of the variables in the interaction term is exogenously determined. This assumption holds plausibly for temperature shocks, as countries can hardly influence their own weather or reallocate because of it. Therefore, even if some channels could be endogenous with respect to temperature, I argue that the interaction terms allow for an unbiased interpretation.

I apply the interaction model for economic variables, as they have a plausibly linear effect on the temperature-sovereign risk connection. However, some climate-related variables could have non-linear effects that are critical to certain thresholds. For instance, Burke et al. (2015) show that a country's temperature has a non-linear impact on GDP growth. As the interaction model above will only partially capture such non-linear effects, I follow the literature (e.g. Zhang et al. (2018), Graff Zivin & Neidell (2014)) and estimate a bin-model for all climate-related channels:

$$\Delta SovereignCreditworthiness_{it} = \sum_m \lambda_m TemperatureAnomaly_{it} * Channel_i^m$$

$$+ TemperatureAnomaly_{it} + \delta Precip_{it} + \gamma_i + \gamma_{rt} + \epsilon_{it}$$

$$(21)$$

In this way, a country is grouped into one of m (time-invariant) bins. For instance, a country could be sorted into a bin for cold, mild or warm countries based on its average yearly temperature. In order to avoid multicollinearity, one bin has to be omitted in the regression. The estimated coefficient λ_m yields the effect of a temperature anomaly increase of, for instance, the warm country group relative to the omitted reference group, for example the mild countries. Thereby, group-specific non-linear temperature effects are taken into account.

5.5.1 General Warmness

Figure 5.3 depicts the histogram of every sample country's 1901-2018 temperature average. There is considerable heterogeneity visible in the warm- and coldness between the coldest (Russia, -4.96°C) and the hottest (Senegal, 28.03°C) country. To test the hypothesis if differences in climatic profiles affect the temperature-sovereign risk connection, I construct five bins to group every country into: very cold, cold, mild, warm and very warm.

Figure 5.3: Histogram of average temperature (1901-2018) of panel countries

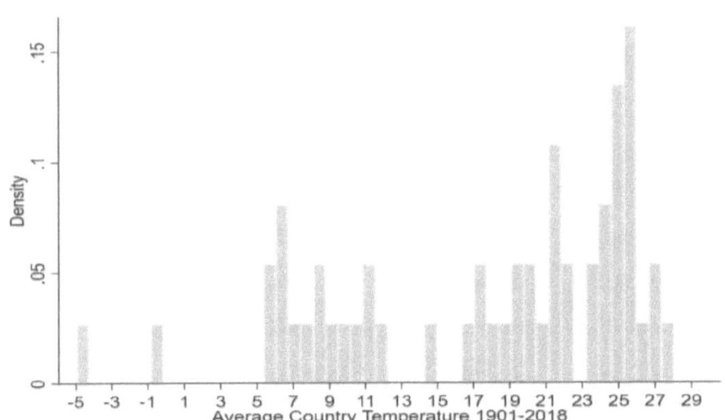

I start by grouping according to percentiles: Countries equal to or below the 20th percentile of average annual temperature (from 1901-2018) are classified as "very cold". Countries in the 21st to the 40th percentile of the sample-wide annual temperature distribution are classified as "cold" and so on. Using this data-driven procedure, I make sure that every bin has the same number of countries.

One drawback of this method is that the differences at the end of the distribution are less sharp. "Warm" countries have an average temperature of 24.36°C while "very warm" countries have only marginally hotter climate averaging 26.25°C. Therefore, for a second procedure, I group according to 5°C-intervals: "Very cold" includes countries with mean 1901-2018 temperatures below 10°C, "cold" ranges between 10°C and 15°C, "mild" between 15°C and 20°C, "warm" between 20°C and 25°C and "very warm" above 25°C. With this procedure, the number of countries in each bin varies. Table 5.7 shows the members of each bin and their mean temperature for both classifications.

I proceed by estimating both bin classifications according to equation (21). I omit the "cold" bin to avoid multicollinearity.[55] Table 5.8 reports the results and Figures 5.4 and 5.5 depict the coefficients. I find that the interaction of the "very warm" category and temperature anomalies is both times negative and statistically significant at the 5% level. As in Burke et al. (2015), warmer countries seem to suffer more from temperature increases

---

[55]I only interact with the unadjusted historical temperature anomaly variable, as the deviation-adjusted temperature variable already captures countries with low seasonality and warm climate.

Table 5.7: Countries in each percentile- or 5°C-interval-defined climatic bin

Percentile-defined climatic bins	Very Cold	Cold	Mild	Warm	Very Warm
	Belarus	Argentina	Angola	Brazil	Belize
	Chile	Azerbaijan	Bolivia	Colombia	Ghana
	China	Croatia	Ecuador	Costa Rica	Indonesia
	Georgia	Hungary	Iraq	Dominican Republic	Ivory Coast
	Kazakhstan	Lebanon	Jordan	Egypt	Malaysia
	Latvia	Morocco	Mexico	El Salvador	Nigeria
	Lithuania	Romania	Namibia	Gabon	Panama
	Mongolia	Serbia	Pakistan	Guatemala	Philippines
	Poland	South Africa	Peru	India	Senegal
	Russia	Turkey	Tunisia	Jamaica	Venezuela
	Ukraine	Uruguay	Zambia	Vietnam	
Average 1901-2018 annual temperature	5.198°C	13.439°C	20.666°C	24.362°C	26.256°C

5°C-interval-defined bins	Very Cold: ≤10°C	Cold: >10 & ≤15°C	Mild: >15 & ≤20°C	Warm: >20 & ≤25°C	Very Warm: >25°C
	Belarus	Argentina	Jordan	Angola	Belize
	Chile	Azerbaijan	Lebanon	Bolivia	Brazil
	China	Croatia	Morocco	Colombia	Gabon
	Georgia	Serbia	Peru	Costa Rica	Ghana
	Hungary	Turkey	South Africa	Dominican Republic	Indonesia
	Kazakhstan		Tunisia	Ecuador	Ivory Coast
	Latvia		Uruguay	Egypt	Jamaica
	Lithuania			El Salvador	Malaysia
	Mongolia			Guatemala	Nigeria
	Poland			India	Panama
	Romania			Iraq	Philippines
	Russia			Mexico	Senegal
	Ukraine			Namibia	Venezuela
				Pakistan	
				Vietnam	
				Zambia	
Average 1901-2018 annual temperature	5.863°C	11.938°C	18.087°C	22.657°C	25.986°C

than milder tempered countries. For both models, the effect of "very warm" countries holds at least at the 10% level of statistical significance with respect to the cold but also the mild and very cold category and for the 5°C-interval model even towards the "warm" category (see Figures A.5.1 and A.5.2).

Independent of the base category, summing the interaction coefficient of "very warm" countries and the single term coefficient of historical temperature anomalies gives the total size of the effect. For "very warm" countries, I find that a rise in historical temperature anomalies by 1°C, i.e. the estimated global temperature increase since the pre-industrial age, leads to a decline in EMBI returns by 0.432%-points in the percentile- and by 0.464%-points in the 5°C-interval model. These effects correspond to 11.1% or 11.9% of the EMBI returns' standard deviation in the sample. Consequently, for affected countries, a 2°C warming scenario could lead to falling sovereign creditworthiness in an amount of roughly 25% of the recent EMBI standard deviation, although this effect is out-of-sample and subject to climate-related uncertainty. One drawback of the EMBI growth data is that, as a financial market return variable, I cannot attach a dollar value to these effects. Still, the magnitude in terms of percentage points and standard deviation shares is quite substantial. The effect implies sharply rising sovereign borrowing costs for sovereigns that are susceptible to climate change.

Table 5.8: Channels of temperature-sovereign risk connection: general warmness

	(1) ΔEMBI	(2) ΔEMBI
HistoricalTempAnomaly	0.0596	0.0825
	(0.0834)	(0.117)
VeryColdCountry (percentile) × HistoricalTempAnomaly	-0.0731	
	(0.0802)	
ColdCountry (percentile; base category) × HistoricalTempAnomaly	0	
	(0)	
MildCountry (percentile) × HistoricalTempAnomaly	-0.0639	
	(0.114)	
WarmCountry (percentile) × HistoricalTempAnomaly	-0.224	
	(0.174)	
VeryWarmCountry (percentile) × HistoricalTempAnomaly	-0.491**	
	(0.233)	
VeryColdCountry (≤ 10°C) × HistoricalTempAnomaly		-0.0492
		(0.110)
ColdCountry (> 10 & ≤ 15°C; base category) × HistoricalTempAnomaly		0
		(0)
MildCountry (> 15 & ≤ 20°C) × HistoricalTempAnomaly		-0.207
		(0.124)
WarmCountry (> 20 & ≤ 25°C) × HistoricalTempAnomaly		-0.125
		(0.155)
VeryWarmCountry (> 25°C) × HistoricalTempAnomaly		-0.547**
		(0.229)
Precipitation	0.0640	0.00819
	(0.406)	(0.393)
Observations	9,957	9,957
R-squared	0.524	0.524
Country FE	Yes	Yes
Region×MonthYear FE	Yes	Yes
Total "very warm" Country Effect	-0.432	-0.464

This table shows OLS estimation results of a panel of 54 countries from 1994m1 to 2018m12. ΔEMBI are monthly natural log returns of a country's EMBI index. HistoricalTempAnomaly is the difference between monthly temperature of a country and its 1901-1950 temperature average of the same month. Each country is grouped into a bin either according to percentiles or 5°C-intervals (see Table 5.7 for respective countries). One bin is omitted due to multicollinearity (base category). Single terms of the bins are subsumed by time fixed effects. Total "very warm" country effect is the sum of the VeryWarmCountry interaction and the single term of HistoricalTempAnomaly. Standard errors (in parentheses) are clustered at the country level, ***, ** and * indicate statistical significance at the 1%, 5% and 10% level, respectively. See Table A.5.5 for variable definitions and sources.

Figure 5.4: Coefficients estimated in Table 5.8 for climatic bins according to percentiles of average temperature

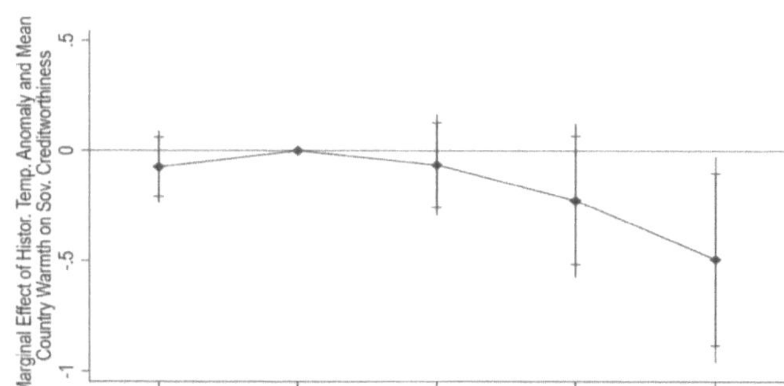

Figure 5.5: Coefficients estimated in Table 5.8 for climatic bins according to 5°C-intervals of average temperature

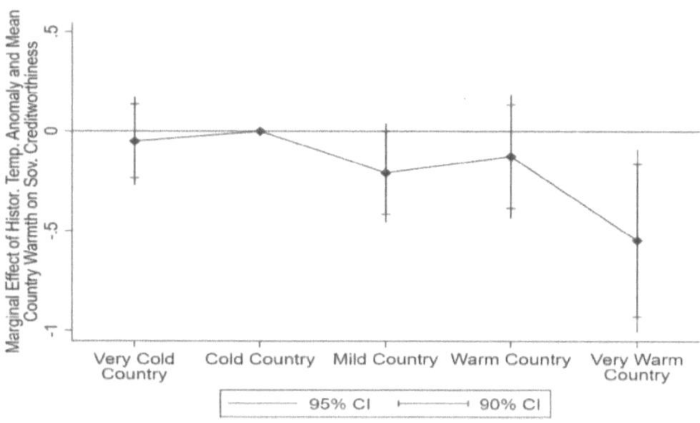

5.5.2 Seasonality

The negative and statistically significant effects of deviation-adjusted temperature anomaly (Table 5.6, columns (4)-(6)) gave already some confirmation that lower seasonality makes a country more susceptible to temperature socks. The evidence from the recent section corroborates this finding, as countries that are warmer on average tend to have lower seasonality (a country's mean temperature and temperature standard deviation correlate at -0.89).

To test the effects of seasonality on temperature-induced sovereign risk more formally, I group each country into one of five seasonality bins. I again sort according to the quantiles (i.e. five percentile groups) of monthly temperature standard deviation (1901-2018) with the grouping shown in Table A.5.1. Table 5.9 and the depicted coefficients in Figure 5.6 confirm the previous results: The coefficient for countries with very low seasonality is negative and statistically significant in its interaction with temperature anomalies. The size is nearly identical to that of the warmest country group in the previous section. In addition, countries in the neighboring group of low seasonality have a negative and statistically significant coefficient at the 10% level. In sum, this evidence suggests that hotter countries where seasons hardly vary in terms of warmness are more sensitive towards rising temperature anomalies with respect to their sovereign creditworthiness.

Table 5.9: Channels of temperature-sovereign risk connection: seasonality

	(1) ΔEMBI
HistoricalTempAnomaly	0.0511
	(0.102)
VeryLow Temp-StdDev (percentile) $\times$ HistoricalTempAnomaly	-0.441**
	(0.199)
Low Temp-StdDev (percentile) $\times$ HistoricalTempAnomaly	-0.493*
	(0.281)
Normal Temp-StdDev (percentile; base category) $\times$ HistoricalTempAnomaly	0
	(0)
High Temp-StdDev (percentile) $\times$ HistoricalTempAnomaly	0.0502
	(0.113)
VeryHigh Temp-StdDev (percentile) $\times$ HistoricalTempAnomaly	-0.107
	(0.111)
Precipitation	-0.0100
	(0.396)
Observations	9,957
R-squared	0.525
Country FE	Yes
Region$\times$MonthYear FE	Yes

This table shows OLS estimation results of a panel of 54 countries from 1994m1 to 2018m12. ΔEMBI are monthly natural log returns of a country's EMBI index. HistoricalTempAnomaly is the difference between monthly temperature of a country and its 1901-1950 temperature average of the same month. Each country is grouped into a bin according to percentiles (see Table A.5.1) of monthly temperature standard deviation from 1901-2018 (Temp-StdDev). One bin is omitted due to multicollinearity (base category). Single terms of the bins are subsumed by time fixed effects. Standard errors (in parentheses) are clustered at the country level, ***, ** and * indicate statistical significance at the 1%, 5% and 10% level, respectively. See Table A.5.5 for variable definitions and sources.

Figure 5.6: Coefficients estimated in Table 5.9 for bins according to percentiles of monthly temperature standard deviation (1901-2018)

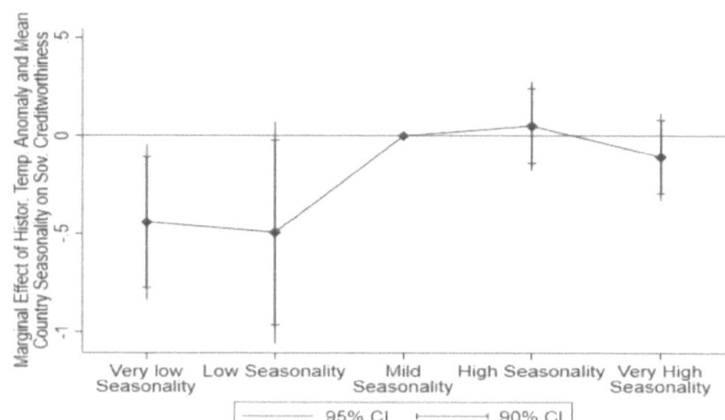

5.5.3 Month and Season Effects

Temperature fluctuations in the previous literature are often on a yearly frequency. I can, instead, exploit the monthly variation in my data to investigate if temperature shocks are different during warmer or colder months. Such a differentiated impact could be plausible: For instance, a warmer summer month could be associated with droughts or declining labor productivity due to extreme heat. On the other hand, a warmer than usual winter month might be beneficial for the economy, as the milder weather lowers heating costs or makes seasonal business cycle fluctuations less severe, for example in the construction sector.

To test these hypotheses, I first re-scale the months of countries in the southern hemisphere to the northern hemisphere classification (that is, the temperature anomaly in Argentina in January is assumed to take place in an adjusted July, February becomes August and so on). I then construct a dummy for each of these adjusted months and repeat the baseline regression by interacting each month with the historical temperature anomaly variable. To avoid multicollinearity, I omit the month of May as it usually approaches the annual temperature average. However, results are not critical towards this choice.

Column (1) in Table 5.10 and Figure 5.7 show the results of this exercise. There seems to be no pattern that would confirm the hypothesis of more severe temperature shocks in summer months. Some months approach statistical significance at the 10% level, however, such findings are not stable and in general sensitive to the omitted base category. For instance, in column (2) I omit December to see potential summer-effects more directly, which leads to changes in signs and significance levels for several coefficients.

Lastly, I interact temperature anomaly shocks with the respective season. This specification does not require a re-scaling of months in southern hemisphere countries (e.g. summer is June-August in northern and December-February in southern hemisphere). Choosing autumn as a base, column (3) of Table 5.10 and Figure 5.8 demonstrate once again that there seem to be no statistically significant effects during certain seasons regarding the temperature-sovereign risk connection. These results lead to the conclusion that overall country warmness rather than within year temperature variation is what matters for the temperature sensitivity of a country's sovereign creditworthiness.

Figure 5.7: Coefficients estimated in Table 5.10 column (1) for month effects (adjusted for southern hemisphere)

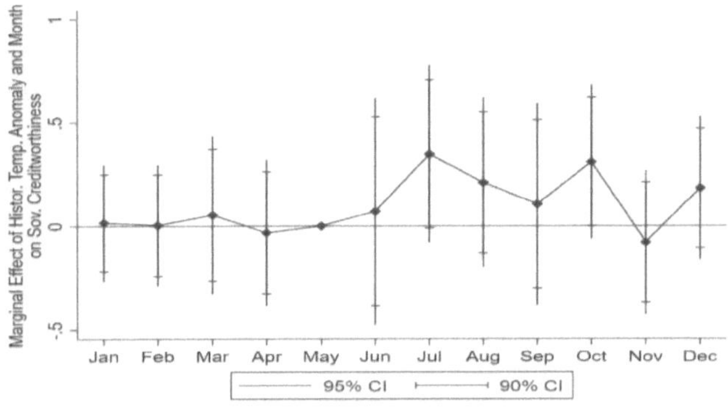

Figure 5.8: Coefficients estimated in Table 5.10 column (3) for season effects

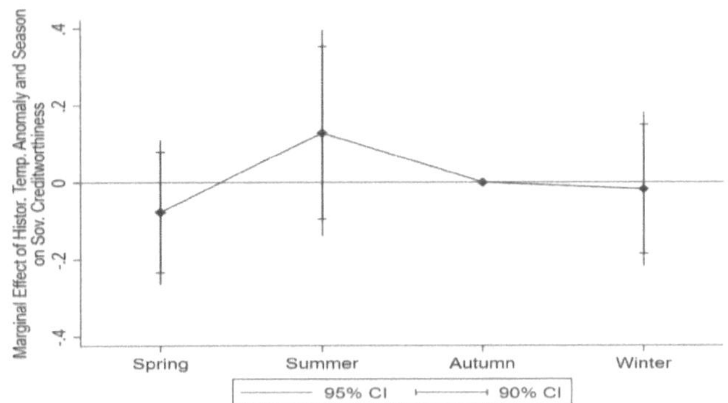

Table 5.10: Channels of temperature-sovereign risk connection: month and season effects

	(1) ΔEMBI	(2) ΔEMBI	(3) ΔEMBI
HistoricalTempAnomaly	-0.0939	0.0890	-0.00309
	(0.124)	(0.106)	(0.102)
January (adjusted) $\times$ HistoricalTempAnomaly	0.0155	-0.167	
	(0.140)	(0.108)	
February (adjusted) $\times$ HistoricalTempAnomaly	0.00336	-0.179	
	(0.146)	(0.137)	
March (adjusted) $\times$ HistoricalTempAnomaly	0.0532	-0.130	
	(0.190)	(0.136)	
April (adjusted) $\times$ HistoricalTempAnomaly	-0.0325	-0.215	
	(0.176)	(0.181)	
May (adjusted; base category in (1)) $\times$ HistoricalTempAnomaly	0	-0.183	
	(0)	(0.173)	
June (adjusted) $\times$ HistoricalTempAnomaly	0.0704	-0.112	
	(0.273)	(0.220)	
July (adjusted) $\times$ HistoricalTempAnomaly	0.347	0.165	
	(0.213)	(0.186)	
August (adjusted) $\times$ HistoricalTempAnomaly	0.210	0.0271	
	(0.204)	(0.160)	
September (adjusted) $\times$ HistoricalTempAnomaly	0.105	-0.0779	
	(0.243)	(0.238)	
October (adjusted) $\times$ HistoricalTempAnomaly	0.309	0.127	
	(0.186)	(0.159)	
November (adjusted) $\times$ HistoricalTempAnomaly	-0.0798	-0.263**	
	(0.173)	(0.109)	
December (adjusted; base category in (2)) $\times$ HistoricalTempAnomaly	0.183	0	
	(0.173)	(0)	
Spring $\times$ HistoricalTempAnomaly			-0.0766
			(0.0935)
Summer $\times$ HistoricalTempAnomaly			0.129
			(0.134)
Autumn (base category) $\times$ HistoricalTempAnomaly			0
			(0)
Winter $\times$ HistoricalTempAnomaly			-0.0175
			(0.0998)
Precipitation	0.362	0.362	0.439
	(0.477)	(0.477)	(0.437)
Observations	9,957	9,957	9,957
R-squared	0.525	0.525	0.524
Single Terms	Yes	Yes	Yes
Country FE	Yes	Yes	Yes
Region$\times$MonthYear FE	Yes	Yes	Yes

This table shows OLS estimation results of a panel of 54 countries from 1994m1 to 2018m12. ΔEMBI are monthly natural log returns of a country's EMBI index. HistoricalTempAnomaly is the difference between monthly temperature of a country and its 1901-1950 temperature average of the same month. In columns (1) and (2), months in the southern hemisphere are adjusted to northern hemisphere scaling (January becomes July and so on). One month or season is omitted due to multicollinearity (base category). The single terms of months or seasons are included in the regression but left out in the table to save space. See Table A.5.2 for the full results. Standard errors (in parentheses) are clustered at the country level, ***, ** and * indicate statistical significance at the 1%, 5% and 10% level, respectively. See Table A.5.5 for variable definitions and sources.

5.5.4 Economic Sector Specialization

In the following, I investigate if countries that are specialized in certain economic sectors are more susceptible to temperature deviating positively from its historical average. For instance, Auffhammer & Schlenker (2014) summarize empirical studies on the tight relationship between agricultural production, weather outcomes and climate change. Furthermore, the literature linking temperature and labor productivity typically finds critical effects in manufacturing sectors (Cachon et al. 2012, Zhang et al. 2018). Lastly, countries specialized in fossil-fuel sectors could see their creditworthiness deteriorate because these industries might no longer have viable business models as climate change intensifies (ECB 2019).

To test the statistical effects of these channels, I interact the temperature anomaly variable with measures for sector specialization as described in equation (20). Though pure temperature anomalies are the primary interest of this specification, I also run the regressions using the deviation-adjusted anomaly measure to emphasize countries with lower seasonality and warmer weather.

Table 5.11 shows the results for agricultural ((1)-(2)), manufacturing ((3)-(4)) and services ((5)-(6)) specialization, all measured by their share on GDP. Negative interaction effects would indicate that higher specialization in a sector leads to more detrimental temperature impacts on sovereign creditworthiness. However, while some coefficients of the interactions with temperature anomaly are negative in sign, none of them are statistically significant at conventional levels. Table 5.25 in the robustness section repeats these interactions with different scaling, for instance the employment instead of the GDP share, but the results stay statistically insignificant. Overall, the gathered evidence does not suggest that countries which are specialized in a certain economic sector are more (or less) susceptible to temperature increases with respect to their sovereign solvency.

Lastly, I interact with total share of oil, gas, coal, mineral and forest rents in relation to GDP (ResourceRentsToGDP) ((7)-(8)). However, there are once again no statistically significant interaction effects for either one of the temperature anomaly variables. Of course, it could still be the case that fossil industries captured in the resource-rent variable will come under stronger pressure in future years and thereby endanger the creditworthiness of their sovereign. Still, such effects seem to be either weak during my estimation period or not connected to the temperature shocks estimated in the model.

	(1) ΔEMBI	(2) ΔEMBI	(3) ΔEMBI	(4) ΔEMBI	(5) ΔEMBI	(6) ΔEMBI	(7) ΔEMBI	(8) ΔEMBI
HistoricalTempAnomaly	0.0388		-0.164		-0.203		0.00517	
	(0.0781)		(0.104)		(0.214)		(0.0587)	
DeviationAdjustedTempAnomaly		-0.290		-0.546*		-4.67e-05		-0.218*
		(0.191)		(0.310)		(0.328)		(0.112)
AgricultureToGDP	0.0188	0.0137						
	(0.0303)	(0.0301)						
HistoricalTempAnomaly × AgricultureToGDP	-0.00828							
	(0.00970)							
DeviationAdjustedTempAnomaly × AgricultureToGDP		0.00615						
		(0.0186)						
ManufacutringToGDP			0.0379	0.0339				
			(0.0314)	(0.0330)				
HistoricalTempAnomaly × ManufacutringToGDP			0.00833					
			(0.00709)					
DeviationAdjustedTempAnomaly × ManufacutringToGDP				0.0192				
				(0.0171)				
ServicesToGDP					-0.0410**	-0.0374**		
					(0.0154)	(0.0148)		
HistoricalTempAnomaly × ServicesToGDP					0.00339			
					(0.00392)			
DeviationAdjustedTempAnomaly × ServicesToGDP						-0.00443		
						(0.00614)		
ResourceRentsToGDP							0.0143	0.0111
							(0.0182)	(0.0182)
HistoricalTempAnomaly × ResourceRentsToGDP							-0.00256	
							(0.00218)	
DeviationAdjustedTempAnomaly × ResourceRentsToGDP								-0.00138
								(0.00742)
Precipitation	0.294	0.111	0.256	0.100	0.296	0.121	0.253	0.0716
	(0.364)	(0.365)	(0.364)	(0.372)	(0.365)	(0.371)	(0.354)	(0.362)
Observations	9,875	9,875	9,599	9,599	9,875	9,875	9,273	9,273
R-squared	0.528	0.528	0.530	0.530	0.529	0.529	0.531	0.531
Country FE	Yes	Yes	Yes	Yes	Yes	Yes	Yes	Yes
Region×MonthYear FE	Yes	Yes	Yes	Yes	Yes	Yes	Yes	Yes

This table shows OLS estimation results of a panel of 54 countries from 1994m1 to 2018m12. ΔEMBI are monthly natural log returns of a country's EMBI index. HistoricalTempAnomaly is the difference between monthly temperature of a country and its 1901-1950 temperature average of the same month. DeviationAdjustedTempAnomaly is the anomaly measure divided by a country's 1901-1950 average of temperature standard deviation. The GDP share of the agriculture sector ((1)-(2)), the manufacturing sector ((3)-(4)) the service sector ((5)-(6)) and the total share of oil, gas, coal, mineral and forest rents in relation to GDP (ResourceRentsToGDP) are used as interaction variables ((7)-(8)). Standard errors (in parentheses) are clustered at the country level, ***, ** and * indicate statistical significance at the 1%, 5% and 10% level, respectively. See Table A.5.5 for variable definitions and sources.

5.5.5 Institutions

The subsequent section investigates if the quality of a country's institutions differentiates the effect of temperature increases on sovereign risk. Better institutions make sure that countries have a stable political and business environment, low corruption, accountable political leaders and a government that can mobilize investments, provide common goods and respond to market failures or natural disasters. All these factors matter in the context of climate change, for instance if droughts or floods lead to physical damages that require swift government intervention, or if distributional consequences of temperature-induced costs and losses need to be managed efficiently. In sum, better institutional quality could make a country more resilient to the various challenges global warming poses for emerging economies.

In order to test the hypothesis of institutions driving the temperature-sovereign connection, I interact both temperature anomaly versions with a range of institutional measures. My main interest lies once again in the raw temperature anomaly measure as it proxies global warming directly and is more straightforward to interpret, but I will also post results for the deviation-adjusted version. The first set of interactions are with the World Bank's institutional measures for the quality of a country's rule of law (Table 5.12, columns (1)-(2)) and its control of corruption (columns (3)-(4)) which capture most of all business and legal aspects of institutions. I continue with interactions measuring the impact of political rights (columns (5)-(6)) and civil liberties (columns (7)-(8)) by Freedom House to see if free elections, freedom of speech and other politically- and societal-related aspects play a role. Next, I analyze the amount of income redistribution from before to after taxes (Table 5.13, columns (1)-(2)) from Solt (2019) to see if more equitable countries that redistribute a larger share of their income, thereby potentially taxing elites and lowering poverty, react differently to temperature shocks. Lastly, I use the Polity2 index (columns (3)-(4)) and its components from the Center for Systemic Peace that show which governments are more democratic (columns (5)-(6)) and hence less authoritarian (columns (7)-(8)).

Table 5.12: Channels of temperature-sovereign risk connection: institutional quality (1)

	(1) ΔEMBI	(2) ΔEMBI	(3) ΔEMBI	(4) ΔEMBI	(5) ΔEMBI	(6) ΔEMBI	(7) ΔEMBI	(8) ΔEMBI
HistoricalTempAnomaly	-0.209**		-0.166**		0.137*		0.0984	
	(0.0873)		(0.0816)		(0.0717)		(0.0747)	
DeviationAdjustedTempAnomaly		-0.584***		-0.572***		0.173		-0.0127
		(0.141)		(0.183)		(0.216)		(0.146)
RuleOfLaw	-0.00507	-0.00556						
	(0.00497)	(0.00453)						
HistoricalTempAnomaly $\times$ RuleOfLaw	0.00396***							
	(0.00123)							
DeviationAdjustedTempAnomaly $\times$ RuleOfLaw		0.00802***						
		(0.00261)						
ControlOfCorruption			-0.00494	-0.00506				
			(0.00497)	(0.00511)				
HistoricalTempAnomaly $\times$ ControlOfCorruption			0.00304***					
			(0.00110)					
DeviationAdjustedTempAnomaly $\times$ ControlOfCorruption				0.00728**				
				(0.00348)				
CivilLiberties					0.149*	0.177**		
					(0.0771)	(0.0861)		
HistoricalTempAnomaly $\times$ CivilLiberties					-0.0454***			
					(0.0152)			
DeviationAdjustedTempAnomaly $\times$ CivilLiberties						-0.124**		
						(0.0526)		
PoliticalRights							0.0230	0.0245
							(0.0538)	(0.0551)
HistoricalTempAnomaly $\times$ PoliticalRights							-0.0324**	
							(0.0125)	
DeviationAdjustedTempAnomaly $\times$ PoliticalRights								-0.0749*
								(0.0397)
Precipitation	0.275	0.114	0.293	0.126	0.218	0.00310	0.231	0.0241
	(0.374)	(0.387)	(0.374)	(0.388)	(0.371)	(0.383)	(0.368)	(0.374)
Observations	9,688	9,688	9,688	9,688	9,929	9,929	9,940	9,940
R-squared	0.502	0.502	0.502	0.502	0.524	0.525	0.524	0.524
Country FE	Yes	Yes	Yes	Yes	Yes	Yes	Yes	Yes
Region$\times$MonthYear FE	Yes	Yes	Yes	Yes	Yes	Yes	Yes	Yes

This table shows OLS estimation results of a panel of 54 countries from 1994m1 to 2018m12. ΔEMBI are monthly natural log returns of a country's EMBI index. HistoricalTempAnomaly is the difference between monthly temperature of a country and its 1901-1950 temperature average of the same month. DeviationAdjustedTempAnomaly is the anomaly measure divided by a country's 1901-1950 average of temperature standard deviation. The rule of law index ((1)-(2)), the control of corruption index ((3)-(4)), the civil liberties index ((5)-(6)) and the political rights index ((7)-(8)) are used as interaction variables. Standard errors (in parentheses) are clustered at the country level, ***, ** and * indicate statistical significance at the 1%, 5% and 10% level, respectively. See Table A.5.5 for variable definitions and sources.

Table 5.13: Channels of temperature-sovereign risk connection: institutional quality (2)

	(1) ΔEMBI	(2) ΔEMBI	(3) ΔEMBI	(4) ΔEMBI	(5) ΔEMBI	(6) ΔEMBI	(7) ΔEMBI	(8) ΔEMBI
HistoricalTempAnomaly	-0.0798 (0.0595)		-0.0525 (0.0453)		-0.118** (0.0530)		0.0503 (0.0548)	
DeviationAdjustedTempAnomaly		-0.292*** (0.0880)		-0.373*** (0.117)		-0.532*** (0.164)		-0.144** (0.0617)
IncomeRedistribution	0.0209 (0.0945)	0.0150 (0.0965)						
HistoricalTempAnomaly × IncomeRedistribution	0.00499*** (0.00152)							
DeviationAdjustedTempAnomaly × IncomeRedistribution		0.0163 (0.0138)						
Polity2			-0.0120 (0.0167)	-0.0171 (0.0199)				
HistoricalTempAnomaly × Polity2			0.0115** (0.00448)					
DeviationAdjustedTempAnomaly × Polity2				0.0301** (0.0142)				
DemocraticGovernments					-0.0274 (0.0285)	-0.0364 (0.0320)		
HistoricalTempAnomaly × DemocraticGovernments					0.0196** (0.00736)			
DeviationAdjustedTempAnomaly × DemocraticGovernments						0.0521** (0.0217)		
AuthoritarianGovernments							0.0449 (0.0373)	0.0596 (0.0481)
HistoricalTempAnomaly × AuthoritarianGovernments							-0.0292*** (0.0101)	
DeviationAdjustedTempAnomaly × AuthoritarianGovernments								-0.0775** (0.0321)
Precipitation	0.446 (0.411)	0.333 (0.421)	0.322 (0.383)	0.140 (0.382)	0.315 (0.392)	0.133 (0.389)	0.336 (0.389)	0.154 (0.386)
Observations	8,308	8,308	9,655	9,655	9,517	9,517	9,517	9,517
R-squared	0.539	0.539	0.542	0.542	0.548	0.548	0.548	0.548
Country FE	Yes	Yes	Yes	Yes	Yes	Yes	Yes	Yes
Region×MonthYear FE	Yes	Yes	Yes	Yes	Yes	Yes	Yes	Yes

This table shows OLS estimation results of a panel of 54 countries from 1994m1 to 2018m12. ΔEMBI are monthly natural log returns of a country's EMBI index. HistoricalTempAnomaly is the difference between monthly temperature of a country and its 1901-1950 temperature average of the same month. DeviationAdjustedTempAnomaly is the anomaly measure divided by a country's 1901-1950 average of temperature standard deviation. The amount of income redistribution from before to after taxes (i.e. the Gini difference from before to after taxes) ((1)-(2)), the Polity2 index ((3)-(4)), the democratic government index ((5)-(6)) and the authoritarian government index ((7)-(8)) are used as interaction variables. Standard errors (in parentheses) are clustered at the country level, ***, ** and * indicate statistical significance at the 1%, 5% and 10% level, respectively. See Table A.5.5 for variable definitions and sources.

Simply put, I find strong and robust evidence for all of these channels. Interactions with institutional variables for which higher values indicate better quality (rule of law, control of corruption, income redistribution, polity2, democratic governments) enter with positive, while measures which are indexed so that higher values imply lower quality (civil liberties, political rights, authoritarian governments) carry negative signs in all cases. For the pure temperature anomaly measure, all interactions are at statistical significance levels of 1% or 5%. In the case of the deviation-adjusted measure, the coefficients are slightly weaker in their statistical significance but significant at conventional levels except for the income redistribution interaction.

Concerning economic sizes, having a value in the rule of law index (which ranges between 0 and 100) at the 10th percentile (15.35) leads to a marginal temperature anomaly effect of -0.148 ($= -0.209 + 0.00396 * 15.35$) which is statistically significant at the 5% level. At this index level, a 1°C increase in temperatures leads to a reduction in EMBI returns by -0.148%-points which is 3.8% of EMBI return standard deviation. While this effect is of a smaller magnitude compared to the "very warm" country coefficient, it still captures a non-negligible variation in EMBI returns. In contrast, having a rule of law index at the 90th percentile (69.23) leads to marginal temperature effect of 0.0654 that is statistically insignificant.

I expand the analysis to investigate if the results also hold for climate-related institutions. To this end, I draw data from the Notre Dame Global Adaption Initiative, which publishes the Notre Dame Global Adaption Index (ND-GAIN). This index takes both the climate-related adaptive readiness of a country as well as its physical and institutional vulnerability towards global warming into account. For instance, the index has a readiness component that covers the economic, governance and social-related institutions of a country that can provide resilience towards damages from climate change. The vulnerability component measures physical and topographical exposure risks and the dependency on climate-sensitive sectors. Theoretically, both the readiness and vulnerability component could affect sovereign risk in its interaction with rising temperatures.

Column (1) of Table 5.14 reports the results of the overall ND-GAIN interacted with temperature anomalies. I obtain a positive coefficient that is statistically significant closely before the 5% level, indicating that countries with stronger climate-related institutions suffer significantly less from rising temperature than less well-prepared countries. Results for the interaction between ND-GAIN and the deviation-adjusted temperature are similarly positive and significant just before the 5% level of statistical significance, suggesting again that countries with lower ND-GAIN scores suffer significant negative temperature shocks on their sovereign creditworthiness level.

Interactions with the readiness component (columns (3)-(4)) and the vulnerability component (columns (5)-(6)) of the ND-GAIN index reveal that the readiness part is driving the results. The corresponding interactions are positive and statistically significant at the 1% level, while the vulnerability interactions are statistically insignificant. This finding is in line with previous results, as the vulnerability component measures the dependency on climate-vulnerable sectors, which were shown to be largely unrelated to the temperature-sovereign risk relationship in the previous section. On the other hand, the readiness component captures climate-related governance factors that correlate positively with the previous measures of institutional quality. The high statistical significance could suggest that climate-related institutional features, like disaster protection or frameworks to support investments in adaptive capacities are crucial to deal with rising temperatures.

In sum, this section provides robust evidence that institutions likely influence the connection between rising temperature and sovereign creditworthiness. Countries with lower institutional quality, both in a traditional and in a climate-related context, have so far been hit significantly harder by temperature deviating from its historical levels. This result could suggest that better institutions can make a country more resilient towards the physical damages from climate change. As future global warming will lead to growing damages, transition costs and distributional issues, having stronger institutions to manage these challenges could be a viable strategy in the adaption process towards climate change.

Table 5.14: Channels of temperature-sovereign risk connection: climate-related institutional quality

	(1) ΔEMBI	(2) ΔEMBI	(3) ΔEMBI	(4) ΔEMBI	(5) ΔEMBI	(6) ΔEMBI
HistoricalTempAnomaly	-0.703* (0.354)		-0.506*** (0.179)		0.144 (0.359)	
DeviationAdjustedTempAnomaly		-1.684** (0.750)		-1.239*** (0.322)		-0.523 (0.803)
ND-GAIN	-0.0531 (0.0441)	-0.0551 (0.0398)				
HistoricalTempAnomaly × ND-GAIN	0.0129* (0.00651)					
DeviationAdjustedTempAnomaly × ND-GAIN		0.0301* (0.0152)				
ReadinessIndex			-2.813 (2.044)	-3.190* (1.789)		
HistoricalTempAnomaly × ReadinessIndex			1.081*** (0.367)			
DeviationAdjustedTempAnomaly × ReadinessIndex				2.636*** (0.830)		
VulnerabilityIndex					13.17 (12.11)	13.15 (11.89)
HistoricalTempAnomaly × VulnerabilityIndex					-0.432 (0.913)	
DeviationAdjustedTempAnomaly × VulnerabilityIndex						0.614 (1.903)
Precipitation	0.301 (0.373)	0.193 (0.390)	0.291 (0.372)	0.174 (0.395)	0.362 (0.366)	0.180 (0.381)
Observations	9,842	9,842	9,842	9,842	9,842	9,842
R-squared	0.510	0.511	0.510	0.511	0.510	0.510
Country FE	Yes	Yes	Yes	Yes	Yes	Yes
Region×MonthYear FE	Yes	Yes	Yes	Yes	Yes	Yes

This table shows OLS estimation results of a panel of 54 countries from 1994m1 to 2018m12. ΔEMBI are monthly natural log returns of a country's EMBI index. HistoricalTempAnomaly is the difference between monthly temperature of a country and its 1901-1950 temperature average of the same month. DeviationAdjustedTempAnomaly is the anomaly measure divided by a country's 1901-1950 average of temperature standard deviation. The Notre Dame Global Adaption Index (ND-GAIN) ((1)-(2)), the readiness component of the ND-GAIN ((3)-(4)) and vulnerability component of the ND-GAIN ((5)-(6)) are used as interaction variables. Standard errors (in parentheses) are clustered at the country level, ***, ** and * indicate statistical significance at the 1%, 5% and 10% level, respectively. See Table A.5.5 for variable definitions and sources.

5.5.6 Combining relevant Channels

A channel that could be related to the impact of institutional quality is economic development. Therefore, I interact temperature anomalies with a country's GDP per capita. Column (1) in Table 5.15 confirms that the level of economic development matters, in that poorer countries' sovereign creditworthiness is statistically significantly stronger damaged by a temperature shock than those of economically more developed countries.

However, it could be the case that poorer countries have larger susceptibility to rising temperatures because they tend to have worse institutions. It could also be the other way around, and the effect of worse institutions only works through the associated lower level of economic development. More broadly, the vulnerability of the warmest countries uncovered in Section 5.5.1 could also be interrelated with institutions and development. For instance, Easterly & Levine (2003) show that countries in tropical climate zones tend to develop worse institutions which lowers their economic progress (see also Sachs (2001)). Indeed, annual average temperatures and the rule of law index correlate negatively in the sample (-0.165), indicating that warmer countries tend to have worse institutions.

A possible if not perfect way to test which channels ultimately matter for the temperature-sovereign connection is to combine all relevant interactions in a single model. I start by adding the interaction of temperature anomalies and the rule of law index, as one of the institutional variables, to the model with interacted GDP per capita (column (2)). While the interaction coefficient for rule of law remains statistically significant and of similar size than in Table 5.12, the GDP per capita interaction with temperature decreases in size and becomes statistically insignificant. This finding provides some confirmation that the effect of lower economic development on the temperature-sovereign relationship is mostly driven by the fact that poorer countries tend to have worse institutions.

In column (3), I add the deviation-adjusted temperature variable to the specification of column (2) to capture temperature shocks in warmer and less seasonal countries. The variable remains negative and statistically significant in column (3), but the interacted rule of law coefficient also stays stable and significant. GDP per capita remains statistically insignificant. This result suggests that, even after controlling for temperature shocks in warmer countries, institutional quality can still cushion the impact of a temperature shock to a significant degree. This finding is confirmed in column (4) in which I replace the deviation-adjusted temperature measure with the five bins representing very cold, cold, mild, warm and very warm countries according to 5°C-intervals. Leaving out the cold country bin, I find that both the interaction of temperature anomaly with the very warm country bin and with the rule of law index continue to stay statistically significant and similar in size than before.

While the long-run effects of climate on institutional development are difficult to entangle and beyond the scope of this paper, the fact that the impact of institutions on the temperature-sovereign connection continues to hold even after controlling for the warmness of a country shows that the institution-channel does not work purely through the climate-channel. In that sense, policy makers, independent of the warmness of their country, have an incentive to improve institutional quality, as it can cushion the impact of rising temperatures on their sovereign risk level.

Table 5.15: Channels of temperature-sovereign risk connection: combining relevant channels

	(1) ΔEMBI	(2) ΔEMBI	(3) ΔEMBI	(4) ΔEMBI
HistoricalTempAnomaly	-0.145*	-0.233***	-0.151*	-0.137
	(0.0764)	(0.0845)	(0.0893)	(0.137)
GDPPerCapita	-5.83e-05	-4.19e-05	-4.04e-05	-3.57e-05
	(4.14e-05)	(4.06e-05)	(4.03e-05)	(3.98e-05)
HistoricalTempAnomaly	1.70e-05**	7.21e-06	5.97e-06	2.71e-06
× GDPPerCapita	(6.54e-06)	(7.80e-06)	(7.54e-06)	(6.80e-06)
RuleOfLaw		-0.00315	-0.00235	-0.00414
		(0.00477)	(0.00474)	(0.00467)
HistoricalTempAnomaly × RuleOfLaw		0.00326**	0.00317**	0.00381***
		(0.00149)	(0.00138)	(0.00133)
DeviationAdjustedTempAnomaly			-0.314***	
			(0.107)	
VeryColdCountry ($\leq 10°$C)				-0.0508
× HistoricalTempAnomaly				(0.101)
ColdCountry (> 10 & $\leq 15°$C; base category)				0
× HistoricalTempAnomaly				(0)
MildCountry (> 15 & $\leq 20°$C)c				-0.217
× HistoricalTempAnomaly				(0.131)
WarmCountry (> 20 & $\leq 25°$C)				-0.0422
× HistoricalTempAnomaly				(0.158)
VeryWarmCountry ($> 25°$C)				-0.539**
× HistoricalTempAnomaly				(0.237)
Precipitation	0.194	0.267	0.0751	0.0683
	(0.367)	(0.375)	(0.387)	(0.392)
Observations	9,957	9,688	9,688	9,688
R-squared	0.524	0.502	0.502	0.502
Country FE	Yes	Yes	Yes	Yes
Region×MonthYear FE	Yes	Yes	Yes	Yes

This table shows OLS estimation results of a panel of 54 countries from 1994m1 to 2018m12. ΔEMBI are monthly natural log returns of a country's EMBI index. HistoricalTempAnomaly is the difference between monthly temperature of a country and its 1901-1950 temperature average of the same month. DeviationAdjustedTempAnomaly is the anomaly measure divided by a country's 1901-1950 average of temperature standard deviation. GDP per capita ((1)-(4)), rule of law ((2)-(4)) and country warmness ((4)) are used as interaction variables. Standard errors (in parentheses) are clustered at the country level, ***, ** and * indicate statistical significance at the 1%, 5% and 10% level, respectively. See Table A.5.5 for variable definitions and sources.

5.6 Robustness Tests

5.6.1 Changing the Fixed Effects Specification

In order to conduct robustness checks, I repeat those specifications that yielded the most decisive results in the previous sections. These include the deviation-adjusted temperature variable (Table 5.6, column (6)), the bin-regression analyzing the warmness of countries using $5°C$-intervals (Table 5.8, column (2)) and the interaction with institutional characteristics from which I choose the rule of law index (Table 5.12, column (1)). For the bin-regression, I omit the "cold" country category because it provides a distinctive comparison group to the "very warm" country group. However, Figures A.5.1 and A.5.2 show that the general effect also holds for omitting other groups. More importantly, I am interested in the total effect of the "very warm" country group interaction, which is independent of the omitted bin.

I start by changing the fixed effects setting for each of these three specifications. First, I deconstruct the interaction of region and year-month fixed effects and instead only include year-month time effects, thus omitting the regional component (Table 5.16, columns (1)-(3)). Second, I re-include the region times month-year effects and in addition interact the country fixed effects with a year time fixed effect (columns (4)-(6)). Though I am not aware of a paper in the relevant literature using such a country-year effect, the interaction controls for time-fixed differences between countries within each year. Lastly, I control for region times month-year and additional country times quarter fixed effects (columns (7)-(9)). The latter interaction absorbs seasonal differences that vary over each quarter.

In sum, the main results of the paper stay intact for each of these modifications. Dropping the region fixed effects only marginally changes the coefficients. The country by year fixed effects, in contrast, reduce the statistical significance of the deviation-adjusted temperature anomaly to the 10% level, and also lower the total effect of very warm countries from -0.464 in the baseline to -0.320 in column (5). Still, this specification is unusual in the literature and the overall direction of the results is the same as before. Interacting the country with quarter fixed effects yields stable and even slightly stronger results than in the baseline.

Table 5.16: Robustness tests: changing fixed effects specification

	(1) ΔEMBI	(2) ΔEMBI	(3) ΔEMBI	(4) ΔEMBI	(5) ΔEMBI	(6) ΔEMBI	(7) ΔEMBI	(8) ΔEMBI	(9) ΔEMBI
HistoricalTempAnomaly		0.0125 (0.0795)	-0.222** (0.0830)		0.0599 (0.112)	-0.194** (0.0952)		0.0913 (0.113)	-0.241*** (0.0852)
DeviationAdjustedTempAnomaly	-0.237*** (0.0743)			-0.179* (0.106)			-0.238*** (0.0880)		
VeryColdCountry × HistoricalTempAnomaly		-0.0482 (0.0918)			-0.0355 (0.104)			-0.0470 (0.110)	
ColdCountry (base category) × HistoricalTempAnomaly		0 (0)			0 (0)			0 (0)	
MildCountry × HistoricalTempAnomaly		-0.120 (0.0994)			-0.176 (0.118)			-0.234* (0.117)	
WarmCountry × HistoricalTempAnomaly		-0.0984 (0.118)			-0.114 (0.146)			-0.196 (0.156)	
VeryWarmCountry × HistoricalTempAnomaly		-0.428** (0.165)			-0.380** (0.185)			-0.562** (0.249)	
RuleOfLaw			-0.00366 (0.00499)						-0.00562 (0.00500)
HistoricalTempAnomaly × RuleOfLaw			0.00346** (0.00148)			0.00358*** (0.00126)			0.00457*** (0.00117)
Precipitation	-0.0832 (0.440)	-0.104 (0.443)	0.286 (0.411)	0.0590 (0.408)	0.0555 (0.404)	0.234 (0.380)	0.481 (0.672)	0.428 (0.690)	0.776 (0.668)
Observations	10,006	10,006	9,699	9,951	9,951	9,682	9,957	9,957	9,688
R-squared	0.418	0.418	0.395	0.571	0.572	0.552	0.532	0.532	0.510
Country FE	Yes	Yes	Yes	No	No	No	No	No	No
MonthYear FE	Yes	Yes	Yes	No	No	No	No	No	No
Region×MonthYear FE	No	No	No	Yes	Yes	Yes	Yes	Yes	Yes
Country×Year FE	No	No	No	Yes	Yes	Yes	No	No	No
Country×Quarter FE	No	No	No	No	No	No	Yes	Yes	Yes
Total "very warm" Country Effect		-0.415			-0.320			-0.471	

This table shows robustness checks for the deviation-adjusted temperature variable (Table 5.6, column (6)), the bin-regression analyzing the warmness of countries using 5°C-intervals (Table 5.8, column (2)) and the interaction with institutional characteristics (rule of law index, Table 5.12, column (1)). The total "very warm" country effect is derived by adding the coefficients of the "very warm" interaction effect and the single term of HistoricalTempAnomaly. I change fixed effects as described in the table. Standard errors (in parentheses) are clustered at the country level, ***, ** and * indicate statistical significance at the 1%, 5% and 10% level, respectively. See Table A.5.5 for variable definitions and sources.

5.6.2 Changing the Dependent Variable

Next, I test if the main results hold when using a different dependent variable. All specifications so far used monthly returns of the EMBI index. A natural alternative for this measure are differences in the EMBI spread instead of the index level.

Table 5.17 repeats the three main regressions using monthly first differences of EMBI spreads as the dependent variable (columns (1)-(3)). All results continue to stay statistically significant if on somewhat lower levels. The coefficient signs are now reversed as rising EMBI spread changes indicate lower sovereign creditworthiness. Regarding the economic magnitude, an increase of 1°C of the anomaly measure in "very warm" countries leads to a 9.62 basis-point increase in EMBI spread changes. This effect is 11.98% of the overall EMBI spread change standard deviation (80.34) and thus extremely close to the 11.9% obtained for the EMBI index returns.

In order to investigate the validity of the results for a different variable than the EMBIs, I collect sovereign CDS data. However, this data is only available since roughly 2008 and only for 37 of the 54 panel countries. With these limitations in mind, I construct changes in the CDS spread the same way as with the EMBI spread, i.e. I take first differences, set zero returns to missing and winsorize at the 1st and 99th percentile. I use CDS spread changes as a new dependent variable in columns (4)-(5). I do not estimate the regression using the temperature bin interactions because the grouping process is significantly biased due to the lower number of countries The interaction with the rule of law index is negative and statistically significant at the 5% level, which corroborates the previous results. The deviation-adjusted measure enters positively but is not statistically significant. However, the imprecise estimation could likely be due to the lower number of observations, since the coefficient size is still large. An increase of deviation-adjusted temperature by one standard deviation of the estimation sample (0.627) increases CDS changes by 5.19 points which is 7.3% of the CDS standard deviation (71.05).

Table 5.17: Robustness tests: changing dependent variable

	(1) ΔEMBI Spread	(2) ΔEMBI Spread	(3) ΔEMBI Spread	(4) ΔCDS Spread	(5) ΔCDS Spread
DeviationAdjustedTempAnomaly	3.667*			8.276	
	(1.837)			(5.990)	
HistoricalTempAnomaly		-1.649	4.776**		11.17**
		(2.248)	(2.287)		(4.513)
VeryColdCountry		1.354			
× HistoricalTempAnomaly		(2.182)			
ColdCountry (base category)		0			
× HistoricalTempAnomaly		(0)			
MildCountry		3.301			
× HistoricalTempAnomaly		(2.208)			
WarmCountry		1.766			
× HistoricalTempAnomaly		(2.580)			
VeryWarmCountry		11.27*			
× HistoricalTempAnomaly		(5.734)			
RuleOfLaw			-0.0610		-0.0811
			(0.146)		(0.217)
HistoricalTempAnomaly			-0.0940***		-0.187**
× RuleOfLaw			(0.0350)		(0.0744)
Precipitation	-14.57*	-13.10	-15.86*	19.17	15.03
	(8.247)	(8.715)	(8.030)	(18.99)	(15.71)
Observations	9,610	9,610	9,491	4,277	4,277
R-squared	0.463	0.464	0.456	0.349	0.351
Number of Countries	54	54	54	37	37
Country FE	Yes	Yes	Yes	Yes	Yes
Region× MonthYear FE	Yes	Yes	Yes	Yes	Yes

This table shows robustness checks for the deviation-adjusted temperature variable (Table 5.6, column (6)), the bin-regression analyzing the warmness of countries using 5°C-intervals (Table 5.8, column (2)) and the interaction with institutional characteristics (rule of law index, Table 5.12, column (1)). Columns (1)-(3) use the first difference of the EMBI spread, and columns (4)-(5) the first difference of the CDS spread as a new dependent variable. Standard errors (in parentheses) are clustered at the country level, ***, ** and * indicate statistical significance at the 1%, 5% and 10% level, respectively. See Table A.5.5 for variable definitions and sources.

5.6.3 Changing the Lag Structure

Dell et al. (2012) include up to ten years of lagged temperature shocks into one of their specifications. Though their effects appear to be driven by contemporaneous temperature fluctuations, I also extend my model with twelve months of lagged temperature anomalies.

However, columns (1)-(3) in Table 5.18 reveal that, similar to Dell et al. (2012), contemporaneous shocks are driving the results. In column (1), the current level of deviation-adjusted temperature remains negative and statistically significant while all its lags are statistically insignificant and without a clear trend. In column (2), I interact every temperature bin-category with the temperature anomaly variable and each of its twelve legs (yielding 52 interaction terms). Table 5.18 only shows the interaction of the "very warm" category and contemporaneous temperature to save space. Full results are in Table A.5.3. The "very warm" country bin still has a negative and significant interaction effect, though the statistical significance is slightly lower likely because of the numerous additional interactions. Lagged interactions are again quite noisy and in almost all cases not statistically significant. Column (3) interacts the rule of law index with the twelve lags of temperature anomaly but the non-lagged version remains the only significant coefficient.

5.6.4   Changing the Historical Temperature Average Period

All main specifications have used 1901-1950 as a historical period to build temperature averages over, from which deviations were calculated. I chose 1950 because it is long enough to ensure a representable average (compared to 1930 or 1940) but with sufficient distance to global temperatures starting to increase more measurably (such as 1960 or 1970).

In Tables 5.19 and 5.20, I repeat all three main estimations using 1930, 1940, 1960 or 1970 as endpoints for the historical average period. All coefficients of interest hardly change as a consequence of these adjusted average periods, including the total effect of temperature anomalies in "very warm" countries.

Table 5.18: Robustness tests: introducing lag structure

	(1) ΔEMBI	(2) ΔEMBI	(3) ΔEMBI
DeviationAdjustedTempAnomaly	-0.275**		
	(0.125)		
HistoricalTempAnomaly		0.0474	-0.204**
		(0.109)	(0.0935)
VeryWarmCountry × HistoricalTempAnomaly		-0.569*	
		(0.297)	
RuleOfLaw			-0.00739
			(0.00642)
HistoricalTempAnomaly × RuleOfLaw			0.00369**
			(0.00139)
Precipitation	0.364	0.291	0.282
	(0.404)	(0.396)	(0.391)
L1.DeviationAdjustedTempAnomaly	0.0914		
	(0.0926)		
L2.DeviationAdjustedTempAnomaly	0.0172		
	(0.144)		
L3.DeviationAdjustedTempAnomaly	0.0130		
	(0.143)		
L4.DeviationAdjustedTempAnomaly	-0.0805		
	(0.156)		
L5.DeviationAdjustedTempAnomaly	0.00625		
	(0.0976)		
L6.DeviationAdjustedTempAnomaly	-0.172*		
	(0.103)		
L7.DeviationAdjustedTempAnomaly	-0.121		
	(0.0946)		
L8.DeviationAdjustedTempAnomaly	0.0379		
	(0.120)		
L9.DeviationAdjustedTempAnomaly	0.0750		
	(0.0779)		
L10.DeviationAdjustedTempAnomaly	-0.0195		
	(0.134)		
L11.DeviationAdjustedTempAnomaly	-0.0749		
	(0.0991)		
L12.DeviationAdjustedTempAnomaly	0.177		
	(0.125)		
L1.HistoricalTempAnomaly × RuleOfLaw			0.00118
			(0.00144)
L2.HistoricalTempAnomaly × RuleOfLaw			-0.00171
			(0.00144)
L3.HistoricalTempAnomaly × RuleOfLaw			-0.00102
			(0.00130)
L4.HistoricalTempAnomaly × RuleOfLaw			0.000323
			(0.00108)
L5.HistoricalTempAnomaly × RuleOfLaw			-0.000594
			(0.00195)
L6.HistoricalTempAnomaly × RuleOfLaw			0.00290*
			(0.00153)
L7.HistoricalTempAnomaly × RuleOfLaw			-0.00241
			(0.00174)
L8.HistoricalTempAnomaly × RuleOfLaw			0.00219
			(0.00158)
L9.HistoricalTempAnomaly × RuleOfLaw			0.00121
			(0.00151)
L10.HistoricalTempAnomaly × RuleOfLaw			0.000187
			(0.00134)
L11.HistoricalTempAnomaly × RuleOfLaw			6.45e-05
			(0.00207)
L12.HistoricalTempAnomaly × RuleOfLaw			0.00113
			(0.00147)
Observations	9,842	9,842	9,688
R-squared	0.511	0.514	0.503
Country FE, Region× MonthYear FE	Yes	Yes	Yes
Lag and Single Terms		Yes	Yes

This table shows robustness checks for the deviation-adjusted temperature variable (Table 5.6, column (6)), the bin-regression analyzing the warmness of countries using 5°C-intervals (Table 5.8, column (2)) and the interaction with institutional characteristics (rule of law index, Table 5.12, column (1)). Estimation for column (2) includes all other bin-category interactions ("cold" country as base category). Estimations for columns (2) and (3) also include all lagged single terms and interactions of HistoricalTempAnomaly. See Table A.5.3 for full results. Standard errors (in parentheses) are clustered at the country level, ***, ** and * indicate statistical significance at the 1%, 5% and 10% level, respectively. See Table A.5.5 for variable definitions and sources.

Table 5.19: Robustness tests: changing historical average period (1930, 1940)

	(1) ΔEMBI	(2) ΔEMBI	(3) ΔEMBI	(4) ΔEMBI	(5) ΔEMBI	(6) ΔEMBI
DeviationAdjustedTempAnomaly (1930)	-0.213*** (0.0720)					
HistoricalTempAnomaly (1930)		0.0789 (0.113)	-0.196** (0.0837)			
VeryColdCountry × HistoricalTempAnomaly (1930)		-0.0444 (0.107)				
ColdCountry (base category) × HistoricalTempAnomaly (1930)		0 (0)				
MildCountry × HistoricalTempAnomaly (1930)		-0.193 (0.122)				
WarmCountry × HistoricalTempAnomaly (1930)		-0.112 (0.152)				
VeryWarmCountry × HistoricalTempAnomaly (1930)		-0.531** (0.224)				
RuleOfLaw			-0.00521 (0.00488)			-0.00518 (0.00493)
HistoricalTempAnomaly (1930) × RuleOfLaw			0.00376*** (0.00113)			
DeviationAdjustedTempAnomaly (1940)				-0.218*** (0.0743)		
HistoricalTempAnomaly (1940)					0.0835 (0.116)	-0.201** (0.0850)
VeryColdCountry × HistoricalTempAnomaly (1940)					-0.0453 (0.109)	
ColdCountry (base category) × HistoricalTempAnomaly (1940)					0 (0)	
MildCountry × HistoricalTempAnomaly (1940)					-0.199 (0.122)	
WarmCountry × HistoricalTempAnomaly (1940)					-0.113 (0.155)	
VeryWarmCountry × HistoricalTempAnomaly (1940)					-0.537** (0.225)	
HistoricalTempAnomaly (1940) × RuleOfLaw						0.00392*** (0.00118)
Precipitation	0.0281 (0.380)	-0.0191 (0.403)	0.267 (0.375)	0.0423 (0.377)	0.0147 (0.395)	0.276 (0.374)
Observations	9,957	9,957	9,688	9,957	9,957	9,688
R-squared	0.524	0.524	0.502	0.524	0.524	0.502
Country FE	Yes	Yes	Yes	Yes	Yes	Yes
Region×MonthYear FE	Yes	Yes	Yes	Yes	Yes	Yes
Total "very warm" Country Effect		-0.452			-0.453	

This table shows robustness checks for the deviation-adjusted temperature variable (Table 5.6, column (6)), the bin-regression analyzing the warmness of countries using 5°C-intervals (Table 5.8, column (2)) and the interaction with institutional characteristics (rule of law index, Table 5.12, column (1)). Historical temperature averages are calculated from 1901 to 1930 or 1940 instead of 1950, as shown in the table. Standard errors (in parentheses) are clustered at the country level, ***, ** and * indicate statistical significance at the 1%, 5% and 10% level, respectively. See Table A.5.5 for variable definitions and sources.

Table 5.20: Robustness tests: changing historical average period (1960, 1970)

	(1) ΔEMBI	(2) ΔEMBI	(3) ΔEMBI	(4) ΔEMBI	(5) ΔEMBI	(6) ΔEMBI
DeviationAdjustedTempAnomaly (1960)	-0.244*** (0.0825)					
HistoricalTempAnomaly (1960)		0.0801 (0.116)	-0.215** (0.0877)			
VeryColdCountry × HistoricalTempAnomaly (1960)		-0.0517 (0.109)				
ColdCountry (base category) × HistoricalTempAnomaly (1960)		0 (0)				
MildCountry × HistoricalTempAnomaly (1960)		-0.211* (0.123)				
WarmCountry × HistoricalTempAnomaly (1960)		-0.128 (0.156)				
VeryWarmCountry × HistoricalTempAnomaly (1960)		-0.541** (0.229)				
RuleOfLaw			-0.00505 (0.00498)			-0.00503 (0.00498)
HistoricalTempAnomaly (1960) × RuleOfLaw			0.00401*** (0.00123)			
DeviationAdjustedTempAnomaly (1970)				-0.253*** (0.0864)		
HistoricalTempAnomaly (1970)					0.0757 (0.116)	-0.219** (0.0878)
VeryColdCountry × HistoricalTempAnomaly (1970)					-0.0470 (0.109)	
ColdCountry (base category) × HistoricalTempAnomaly (1970)					0 (0)	
MildCountry × HistoricalTempAnomaly (1970)					-0.214* (0.123)	
WarmCountry × HistoricalTempAnomaly (1970)					-0.122 (0.155)	
VeryWarmCountry × HistoricalTempAnomaly (1970)					-0.553** (0.233)	
HistoricalTempAnomaly (1970) × RuleOfLaw						0.00405*** (0.00124)
Precipitation	0.0384 (0.375)	0.0154 (0.390)	0.272 (0.374)	0.0467 (0.376)	0.0178 (0.389)	0.272 (0.373)
Observations	9,957	9,957	9,688	9,957	9,957	9,688
R-squared	0.524	0.524	0.502	0.524	0.524	0.502
Country FE	Yes	Yes	Yes	Yes	Yes	Yes
Region×MonthYear FE	Yes	Yes	Yes	Yes	Yes	Yes
Total "very warm" Country Effect		-0.461			-0.477	

This table shows robustness checks for the deviation-adjusted temperature variable (Table 5.6, column (6)), the bin-regression analyzing the warmness of countries using 5°C-intervals (Table 5.8, column (2)) and the interaction with institutional characteristics (rule of law index, Table 5.12, column (1)). Historical temperature averages are calculated from 1901 to 1960 or 1970 instead of 1950, as shown in the table. Standard errors (in parentheses) are clustered at the country level, ***, ** and * indicate statistical significance at the 1%, 5% and 10% level, respectively. See Table A.5.5 for variable definitions and sources.

5.6.5 Dropping Countries with lower Data Coverage and larger Landmass

In the main specification, I included all countries with liquid EMBI return data of at least six years. I chose this criterion to manage the trade-off between having a large panel and sufficient observations for each country in the sample. In columns (1)-(3) of Table 5.21, I set the inclusion criterion to ten years (120 months) of liquid EMBI return data. 15 countries in the original sample are affected by this requirement (Angola, Azerbaijan, Belarus, Bolivia, Costa Rica, Guatemala, India, Jordan, Latvia, Lithuania, Mongolia, Namibia, Romania, Senegal, Zambia). I drop these countries and repeat the three main regressions. The number of observations only decreases slightly as a result of this adjustment, and all the main effects retain their statistical significance. The effect of temperature increases in the warmest countries even rises somewhat, in both magnitude and significance.

One further concern I address deals with countries covering a huge landmass. Nations like Russia or China could have several climate zones which makes their temperature average only a rough measure for weather fluctuations. Therefore, I drop the ten countries with the largest landmass from my sample (Russia, China, Brazil, India, Argentina, Kazakhstan, Mexico, Indonesia, Mongolia, Peru) and repeat the main regressions. Columns (4)-(6) reveal that the number of observations now decreases more notably. However, the main results remain broadly intact. Deviation-adjusted temperature shocks even increase, as does the interacted effect of institutions and temperature anomalies. The "very warm" country bin is now marginally insignificant just before the 10% level, perhaps because of the lower number of observations or the changing number of countries in each bin. Nevertheless, the total effect of this group still has the same size as in the main regression (-0.443).

Table 5.21: Robustness tests: dropping countries with lower data coverage and larger landmass

	(1) ΔEMBI	(2) ΔEMBI	(3) ΔEMBI	(4) ΔEMBI	(5) ΔEMBI	(6) ΔEMBI
DeviationAdjustedTempAnomaly	-0.239***			-0.315***		
	(0.0861)			(0.108)		
HistoricalTempAnomaly		0.0994	-0.184*		-0.0587	-0.336***
		(0.131)	(0.0993)		(0.0699)	(0.0875)
VeryColdCountry		-0.0286			0.0559	
× HistoricalTempAnomaly		(0.120)			(0.0898)	
ColdCountry (base category)		0			0	
× HistoricalTempAnomaly		(0)			(0)	
MildCountry		-0.231*			-0.0797	
× HistoricalTempAnomaly		(0.137)			(0.0784)	
WarmCountry		-0.104			-0.0143	
× HistoricalTempAnomaly		(0.166)			(0.120)	
VeryWarmCountry		-0.642***			-0.384	
× HistoricalTempAnomaly		(0.248)			(0.239)	
RuleOfLaw			-0.00479			-0.00261
			(0.00501)			(0.00513)
HistoricalTempAnomaly			0.00381***			0.00514***
× RuleOfLaw			(0.00140)			(0.00138)
Precipitation	0.0319	-0.0355	0.322	-0.0579	-0.0419	0.0934
	(0.418)	(0.445)	(0.414)	(0.522)	(0.539)	(0.523)
Observations	8,746	8,746	8,477	7,641	7,641	7,550
R-squared	0.529	0.529	0.505	0.524	0.524	0.509
Country FE	Yes	Yes	Yes	Yes	Yes	Yes
Region×MonthYear FE	Yes	Yes	Yes	Yes	Yes	Yes
Number of Countries	39	39	39	44	44	44
Total "very warm" Country Effect		-0.543			-0.443	

This table shows robustness checks for the deviation-adjusted temperature variable (Table 5.6, column (6)), the bin-regression analyzing the warmness of countries using 5°C-intervals (Table 5.8, column (2)) and the interaction with institutional characteristics (rule of law index, Table 5.12, column (1)). In columns (1)-(3), all countries with ΔEMBI data of fewer than ten years are dropped. In columns (4)-(6), the ten countries with the largest landmass are dropped. Standard errors (in parentheses) are clustered at the country level, ***, ** and * indicate statistical significance at the 1%, 5% and 10% level, respectively. See Table A.5.5 for variable definitions and sources.

5.6.6 Other Temperature Anomaly Measures

I construct one further measure to detect weather anomaly shocks. This variable is inspired by the fact that not only increases of temperature levels but also of variability are verified as one of the detrimental impacts of climate change (Bathiany et al. 2018). I take the standard deviation of monthly temperature over 12-month rolling windows for every country. This variable captures the volatility of weather over the previous year. I subtract from this measure a country's temperature standard deviation from 1901 to 1950. In this way, similar to the historical temperature anomaly measure, I capture deviations of temperature volatility above its pre-global warming average (HistoricalDeviationAnomaly).

Table 5.22 presents the results for this variable. The historical change in temperature standard deviation is negative and statistically significant as a single variable (column (1)). Its size is similar to the deviation-adjusted temperature shocks, which suggests that periods of more volatile weather can hurt sovereign creditworthiness. However, the interaction coefficients of the variable with both institutional quality and the warmest countries are statistically insignificant (columns (2)-(3)). Though the point estimates are actually comparable to the main regressions or even larger, the effects are imprecisely estimated. This result could suggest that more volatile weather hurts all countries' sovereign bond performance, independent of their climate zone or institutional framework.

Table 5.22: Robustness tests: other temperature anomaly measures

	(1) ΔEMBI	(2) ΔEMBI	(3) ΔEMBI
HistoricalDeviationAnomaly	-0.271**	-0.256*	-0.445***
	(0.121)	(0.152)	(0.155)
VeryColdCountry		-0.0280	
× HistoricalDeviationAnomaly		(0.179)	
ColdCountry (base category)		0	
× HistoricalDeviationAnomaly		(0)	
MildCountry		0.197	
× HistoricalDeviationAnomaly		(0.259)	
WarmCountry		-0.0638	
× HistoricalDeviationAnomaly		(0.385)	
VeryWarmCountry		-0.419	
× HistoricalDeviationAnomaly		(0.439)	
RuleOfLaw			-0.00175
			(0.00544)
HistoricalDeviationAnomaly			0.00395
× RuleOfLaw			(0.00356)
Precipitation	0.209	0.214	0.310
	(0.361)	(0.358)	(0.363)
Observations	9,957	9,957	9,688
R-squared	0.524	0.524	0.502
Country FE	Yes	Yes	Yes
Region×MonthYear FE	Yes	Yes	Yes

This table shows robustness checks for the deviation-adjusted temperature variable (Table 5.6, column (6)), the bin-regression analyzing the warmness of countries using 5°C-intervals (Table 5.8, column (2)) and the interaction with institutional characteristics (rule of law index, Table 5.12, column (1)). HistoricalDeviationAnomaly is a country's standard deviation of temperature over the (rolling) past 12 months minus its 1901-1950 standard deviation of temperature. Standard errors (in parentheses) are clustered at the country level, ***, ** and * indicate statistical significance at the 1%, 5% and 10% level, respectively. See Table A.5.5 for variable definitions and sources.

5.6.7 Analyzing Debt Sustainability

Analyzing the sovereign creditworthiness of emerging economies raises the issue of debt sustainability. Therefore, I interact both temperature variables with measures for debt to GDP ((1)-(2)), the number of years since the last sovereign debt restructuring ((3)-(4)) and the primary net lending of the government scaled to GDP ((5)-(6)) in Table 5.23. Debt to GDP is available for 53 and primary lending for 42 of the panel countries.[56] However, all interactions yield statistically insignificant coefficients, suggesting that temperature shocks do not work primarily through government debt characteristics in their impact on sovereign risk.

5.6.8 Testing for Transition Risks

Though it is, as described in Section 5.2, extremely difficult to differentiate between physical and transition risks in the temperature literature, I conduct a test that could possibly detect transition risks. To this end, I use the Paris Climate Agreement, which was sealed in December 2015, as a transition shock. With the Paris Agreement, almost all countries in the world agreed to limit global warming to well below 2°C. If temperature increases also feature a transition risk component, it could be the case that temperature shocks have stronger impacts on sovereign creditworthiness since the Paris Agreement, because investors are more sensitive towards climate issues.

To test this channel, I interact the three main regressions as well as raw temperature anomalies with a time dummy for the Paris Agreement that is 1 after December 2015. For the temperature anomaly and the deviation-adjusted temperature measure, the Paris dummy does not differentiate the impact of these variables, as the interaction effects are statistically insignificant (Table 5.24, columns (1)-(2)). The results are similar for "very warm" countries and institutions (columns (3)-(4)): The double interaction of temperature and rule of law remains statistically significant and comparable to previous results, whereas the triple interaction with the Paris dummy is small and statistically insignificant. Although this is no

---

[56]For the years since the last debt restructuring, I use the database by Laeven & Valencia (2018) and calculate the number of years since the last sovereign debt restructuring event, including those before the start of my estimation in 1994. A value of zero indicates debt restructuring in the current year. Overall, 32 of my panel countries negotiated at least one sovereign debt restructuring. For the countries without any debt restructuring, I set the value of the corresponding variable to 36, which is the maximum number of years without debt restructuring.

definitive result, it could suggest that temperature shocks are first and foremost a physical risk source, which is largely independent of climate agreements or transition risks.

Table 5.23: Robustness tests: analyzing debt sustainability

	(1) ΔEMBI	(2) ΔEMBI	(3) ΔEMBI	(4) ΔEMBI	(5) ΔEMBI	(6) ΔEMBI
HistoricalTempAnomaly	-0.0567		0.0156		0.00862	
	(0.0573)		(0.100)		(0.0532)	
DebtToGDP	-8.35e-05	0.00171				
	(0.00434)	(0.00537)				
HistoricalTempAnomaly	0.000903					
$\times$ DebtToGDP	(0.00138)					
DeviationAdjustedTempAnomaly		-0.178		-0.141		-0.177**
		(0.159)		(0.116)		(0.0778)
DeviationAdjustedTempAnomaly		-0.00281				
$\times$ DebtToGDP		(0.00239)				
YearsSinceLastSovDebtRestructuring			-0.0288***	-0.0277***		
			(0.00858)	(0.00815)		
HistoricalTempAnomaly			-0.000964			
$\times$ YearsSinceLastSovDebtRestructuring			(0.00304)			
DeviationAdjustedTempAnomaly				-0.00427		
$\times$ YearsSinceLastSovDebtRestructuring				(0.00344)		
PrimaryNetLendingGDP					0.0680***	0.0663***
					(0.0161)	(0.0172)
HistoricalTempAnomaly					-0.00814	
$\times$ PrimaryNetLendingGDP					(0.0113)	
DeviationAdjustedTempAnomaly						-0.00747
$\times$ PrimaryNetLendingGDP						(0.0118)
Precipitation	0.571	0.301	0.222	0.0236	0.387	0.175
	(0.377)	(0.387)	(0.362)	(0.372)	(0.549)	(0.567)
Observations	9,327	9,327	9,957	9,957	7,745	7,745
R-squared	0.509	0.510	0.525	0.525	0.523	0.523
Country FE	Yes	Yes	Yes	Yes	Yes	Yes
Region$\times$MonthYear FE	Yes	Yes	Yes	Yes	Yes	Yes

This table shows OLS estimation results of a panel of 54 countries from 1994m1 to 2018m12. ΔEMBI are monthly natural log returns of a country's EMBI index. HistoricalTempAnomaly is the difference between monthly temperature of a country and its 1901-1950 temperature average of the same month. DeviationAdjustedTempAnomaly is the anomaly measure divided by a country's 1901-1950 average of temperature standard deviation. The debt to GDP ratio ((1)-(2)), number of years since the last sovereign debt restructuring (with maximum value of 36 for countries without any restructuring) ((3)-(4)) and government primary net lending ((5)-(6)) are used as interaction variables. Standard errors (in parentheses) are clustered at the country level, ***, ** and * indicate statistical significance at the 1%, 5% and 10% level, respectively. See Table A.5.5 for variable definitions and sources.

Table 5.24: Robustness tests: Paris Agreement as transition shock

	(1) ΔEMBI	(2) ΔEMBI	(3) ΔEMBI	(4) ΔEMBI
HistoricalTempAnomaly	-0.00387		0.0963	-0.211**
	(0.0608)		(0.147)	(0.104)
HistoricalTempAnomaly $\times$ PostParis	-0.0363		-0.0631	0.0255
	(0.0651)		(0.159)	(0.129)
DeviationAdjustedTempAnomaly		-0.246***		
		(0.0873)		
DeviationAdjustedTempAnomaly $\times$ PostParis		0.0522		
		(0.0843)		
HistoricalTempAnomaly $\times$ VeryWarmCountry			-0.608**	
			(0.257)	
VeryWarmCountry $\times$ PostParis			-0.335	
			(0.430)	
HistoricalTempAnomaly $\times$ VeryWarmCountry $\times$ PostParis			0.375	
			(0.340)	
RuleOfLaw				-0.00551
				(0.00504)
HistoricalTempAnomaly $\times$ RuleOfLaw				0.00415***
				(0.00146)
RuleOfLaw $\times$ PostParis				0.00255
				(0.00374)
HistoricalTempAnomaly $\times$ RuleOfLaw $\times$ PostParis				-0.00113
				(0.00207)
Precipitation	0.224	0.0364	0.0261	0.274
	(0.363)	(0.376)	(0.398)	(0.374)
Observations	9,957	9,957	9,957	9,688
R-squared	0.524	0.524	0.525	0.502
Country FE	Yes	Yes	Yes	Yes
Region$\times$ MonthYear FE	Yes	Yes	Yes	Yes
Other Bin Terms			Yes	

This table shows robustness checks for the temperature anomaly measure (Table 5.6, column (3)), the deviation-adjusted temperature variable (Table 5.6, column (6)), the bin-regression analyzing the warmness of countries using 5°C-intervals (Table 5.8, column (2)) and the interaction with institutional characteristics (rule of law index, Table 5.12, column (1)). PostParis is a dummy with value 1 after the Paris Climate Agreement in December 2015. Estimation in column (3) also includes all other bin categories (cold as base category) and respective interactions, see Table A.5.4 for full results. Standard errors (in parentheses) are clustered at the country level, ***, ** and * indicate statistical significance at the 1%, 5% and 10% level, respectively. See Table A.5.5 for variable definitions and sources.

5.6.9 Changing Economic Sector Specialization Measures

Table 5.11 suggested that being specialized in certain economic sectors does not seem to make a country respond differently to temperature shocks with respect to their sovereign creditworthiness. One reason for this result could be in the way I measured sector specialization. Table 5.25 repeats the interactions with different scaling than a sector's GDP share. However, scaling the sectors by employment share or the agricultural sector by land share also yields statistically insignificant results, in line with the main section.[57]

---

[57]There is no data series for the employment share in manufacturing, but I expect industrial sector employment shares to be closely correlated.

Table 5.25: Robustness tests: changing economic sector specialization

	(1) ΔEMBI	(2) ΔEMBI	(3) ΔEMBI	(4) ΔEMBI	(5) ΔEMBI	(6) ΔEMBI	(7) ΔEMBI	(8) ΔEMBI
HistoricalTempAnomaly	-0.0885		0.0627		-0.386*		-0.108	
	(0.0964)		(0.0824)		(0.228)		(0.205)	
DeviationAdjustedTempAnomaly		-0.179		-0.305*		-0.503		0.0826
		(0.251)		(0.153)		(0.387)		(0.424)
AgricultureLandShare	0.0183	0.0164						
	(0.0320)	(0.0316)						
HistoricalTempAnomaly × AgricultureLandShare	-0.00219							
	(0.00144)							
DeviationAdjustedTempAnomaly × AgricultureLandShare		-0.00199						
		(0.00617)						
AgricultureEmploymentShare			0.0391**	0.0363**				
			(0.0167)	(0.0161)				
HistoricalTempAnomaly × AgricultureEmploymentShare			-0.00364					
			(0.00323)					
DeviationAdjustedTempAnomaly × AgricultureEmploymentShare				0.00317				
				(0.00657)				
IndustrialEmploymentShare					-0.0376	-0.0307		
					(0.0290)	(0.0281)		
HistoricalTempAnomaly × IndustrialEmploymentShare					0.0157			
					(0.00955)			
DeviationAdjustedTempAnomaly × IndustrialEmploymentShare						0.0129		
						(0.0174)		
ServicesEmploymentShare							-0.0459*	-0.0427
							(0.0266)	(0.0259)
HistoricalTempAnomaly × ServicesEmploymentShare							0.00171	
							(0.00352)	
DeviationAdjustedTempAnomaly × ServicesEmploymentShare								-0.00551
								(0.00723)
Precipitation	0.298	0.0889	0.211	0.0323	0.174	0.0042	0.222	0.0296
	(0.369)	(0.387)	(0.364)	(0.379)	(0.363)	(0.376)	(0.364)	(0.378)
Observations	8,662	8,662	9,957	9,957	9,957	9,957	9,957	9,957
R-squared	0.529	0.529	0.524	0.524	0.524	0.524	0.524	0.524
Country FE	Yes	Yes	Yes	Yes	Yes	Yes	Yes	Yes
Region×MonthYear FE	Yes	Yes	Yes	Yes	Yes	Yes	Yes	Yes

This table shows OLS estimation results of a panel of 54 countries from 1994m1 to 2018m12. ΔEMBI are monthly natural log returns of a country's EMBI index. HistoricalTempAnomaly is the difference between monthly temperature of a country and its 1901-1950 temperature average of the same month. DeviationAdjustedTempAnomaly is the anomaly measure divided by a country's 1901-1950 average of temperature standard deviation. The land share of the agriculture sector ((1)-(2)), the employment share of the agriculture sector ((3)-(4)), the employment share of the industrial sector ((5)-(6)), and the employment share of the service sector ((7)-(8)) are used as interaction variables. Standard errors (in parentheses) are clustered at the country level, ***, ** and * indicate statistical significance at the 1%, 5% and 10% level, respectively. See Table A.5.5 for variable definitions and sources.

5.7   Conclusion

I extend the literature on temperature fluctuations to finance, specifically the sovereign debt performance of emerging economies. To this end, I collect monthly temperature data since 1901 for 54 emerging markets. For each country, I calculate the temperature deviation of every month from this month's 1901-1950 temperature average. I run my main empirical analysis from 1994m1 to 2018m12, up until this temperature anomaly is on average $0.84°C$, reflecting past climate change trends. In line with previous literature, I argue that rising temperature deviations approximate physical weather and climate damages.

I regress Emerging-Market-Bond-Index returns on temperature anomalies while controlling for established country, time and region fixed-effects. My main result is that the effects of temperature anomalies on the cost of sovereign debt critically hinge on conditioning factors. Temperature deviations lower sovereign bond performance (i.e. increase sovereign risk) significantly for countries that are (i) warmer on average, (ii) less seasonal, (iii) and have lower institutional quality, both in terms of traditional- and climate-related metrics. Importantly, the effects of institutional quality and the warmness of a country on the temperature-sovereign risk connection hold simultaneously, which implies that stronger institutions can improve the resilience of a country towards climate change, independent of its climatic profile.

The economic effects of temperature increases are more than noteworthy. According to my analysis if a country with an average annual temperature above $25°C$ faces a $1°C$ increase in monthly temperature compared to its historical mean, its EMBI returns are lowered by 0.464 percentage points on average. This effect corresponds to 11.9% of the EMBI returns' overall standard deviation. Hence, a $2°C$ global warming scenario could lower EMBI returns of affected countries by roughly a quarter of their overall standard deviation.

This magnitude suggests that, in the absence of climate-adaption strategies, affected countries likely face considerable increases in their sovereign borrowing costs if temperatures continue to rise due to climate change. These results also raise distributional questions: As of 2017, the countries in my panel were responsible for just 36.6% of accumulated historical global CO_2 emissions but posed 66.2% of the global population. Policy action to limit the degree of global warming and to build adaptive capacities through stronger institutional frameworks are therefore called for.

A.5 Appendix to Chapter 5

Figure A.5.1: Coefficients of climatic bins according to percentiles of average temperature with different reference groups

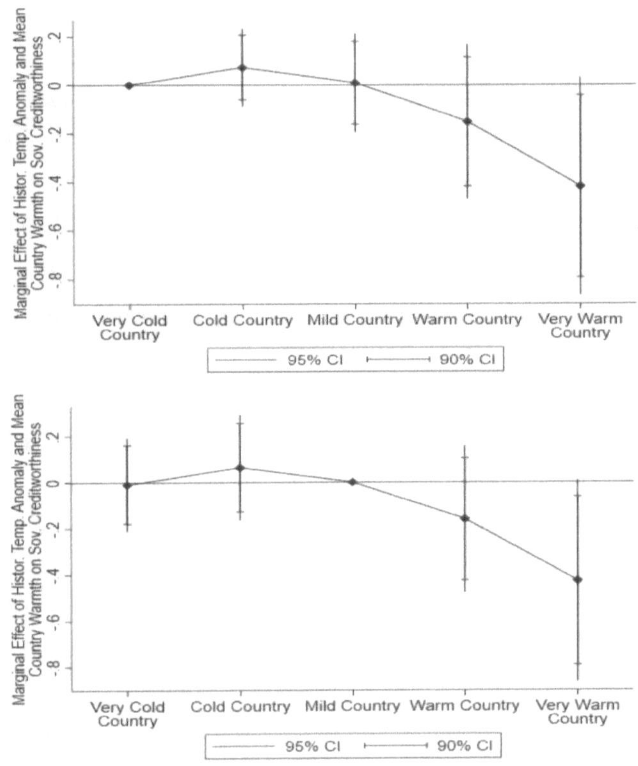

Figure A.5.2: Coefficients of climatic bins according to according to 5°C-intervals of average temperature with different reference groups

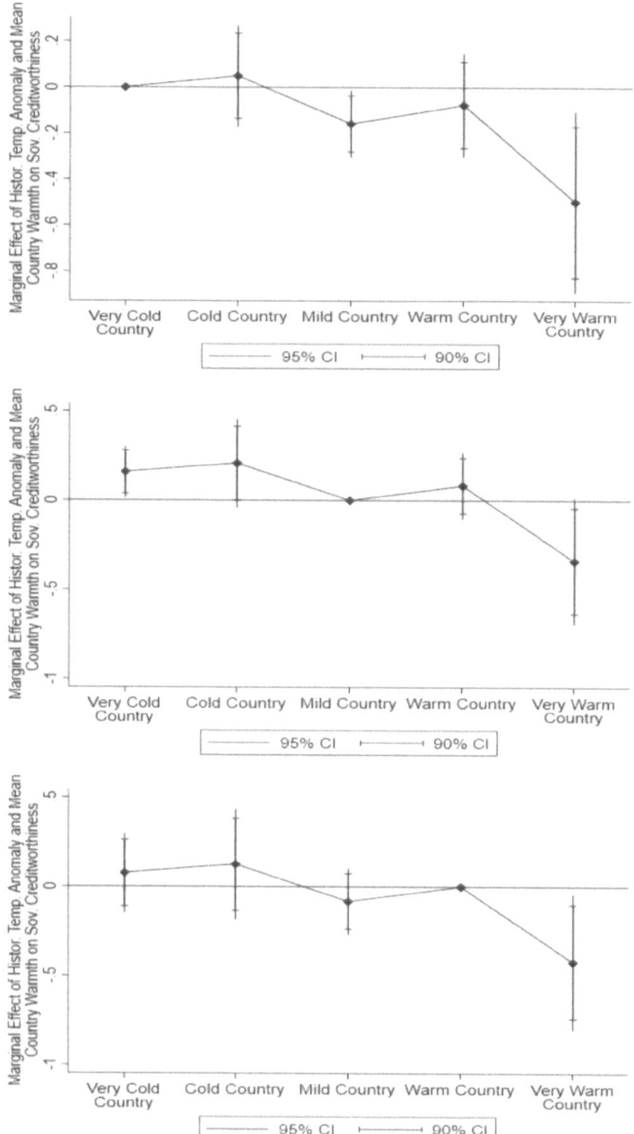

Table A.5.1: Countries in each percentile-defined seasonality bin

	Very low Seasonality	Low Seasonality	Mild Seasonality	High Seasonality	Very high Seasonality
	Brazil	Angola	Argentina	Croatia	Azerbaijan
	Colombia	Belize	Chile	Hungary	Belarus
	Costa Rica	Bolivia	Egypt	Jordan	China
	Ecuador	Dominican Republic	India	Latvia	Georgia
	El Salvador	Gabon	Mexico	Lebanon	Iraq
	Indonesia	Ghana	Morocco	Lithuania	Kazakhstan
	Malaysia	Guatemala	Namibia	Pakistan	Mongolia
	Panama	Ivory Coast	South Africa	Poland	Romania
	Peru	Jamaica	Uruguay	Serbia	Russia
	Philippines	Nigeria	Vietnam	Tunisia	Ukraine
	Venezuela	Senegal	Zambia	Turkey	
Monthly temperature standard deviation 1901-2018	0.681°C	1.52°C	4.07°C	7.50°C	10.36°C

Table A.5.2: Channels of temperature-sovereign risk connection: month and season effects (full table)

	(1) ΔEMBI	(2) ΔEMBI	(3) ΔEMBI
HistoricalTempAnomaly	-0.0939	0.0890	-0.00309
	(0.124)	(0.106)	(0.102)
January (adjusted)	0.124	0.124	
	(0.291)	(0.291)	
January (adjusted) $\times$ HistoricalTempAnomaly	0.0155	-0.167	
	(0.140)	(0.108)	
February (adjusted)	0.401	0.401	
	(0.253)	(0.253)	
February (adjusted) $\times$ HistoricalTempAnomaly	0.00336	-0.179	
	(0.146)	(0.137)	
March (adjusted)	0.175	0.175	
	(0.223)	(0.223)	
March (adjusted) $\times$ HistoricalTempAnomaly	0.0532	-0.130	
	(0.190)	(0.136)	
April (adjusted)	0.317	0.317	
	(0.266)	(0.266)	
April (adjusted) $\times$ HistoricalTempAnomaly	-0.0325	-0.215	
	(0.176)	(0.181)	
May (adjusted; base category in (1))	0	0.399*	
	(0)	(0.206)	
May (adjusted; base category in (1)) $\times$ HistoricalTempAnomaly	0	-0.183	
	(0)	(0.173)	
June (adjusted)	0.414*	0.414*	
	(0.247)	(0.247)	
June (adjusted) $\times$ HistoricalTempAnomaly	0.0704	-0.112	
	(0.273)	(0.220)	
July (adjusted)	0	0	
	(0)	(0)	
July (adjusted) $\times$ HistoricalTempAnomaly	0.347	0.165	
	(0.213)	(0.186)	
August (adjusted)	0	0	
	(0)	(0)	
August (adjusted) $\times$ HistoricalTempAnomaly	0.210	0.0271	
	(0.204)	(0.160)	
September (adjusted)	0	0	
	(0)	(0)	
September (adjusted) $\times$ HistoricalTempAnomaly	0.105	-0.0779	
	(0.243)	(0.238)	
October (adjusted)	0	0	
	(0)	(0)	
October (adjusted) $\times$ HistoricalTempAnomaly	0.309	0.127	
	(0.186)	(0.159)	
November (adjusted)	-0.399*	0	
	(0.206)	(0)	
November (adjusted) $\times$ HistoricalTempAnomaly	-0.0798	-0.263**	
	(0.173)	(0.109)	
December (adjusted; base category in (2))	0	0	
	(0)	(0)	
December (adjusted; base category in (2)) $\times$ HistoricalTempAnomaly	0.183	0	
	(0.173)	(0)	
Spring			0.285**
			(0.127)
Spring $\times$ HistoricalTempAnomaly			-0.0766
			(0.0935)
Summer			-0.0459

	(1)	(2)	(3)
			(0.119)
Summer × HistoricalTempAnomaly			0.129
			(0.134)
Autumn			0
			(0)
Autumn (base category) × HistoricalTempAnomaly			0
			(0)
Winter			0
			(0)
Winter × HistoricalTempAnomaly			-0.0175
			(0.0998)
Precipitation	0.362	0.362	0.439
	(0.477)	(0.477)	(0.437)
Observations	9,957	9,957	9,957
R-squared	0.525	0.525	0.524
Country FE	Yes	Yes	Yes
Region× MonthYear FE	Yes	Yes	Yes

This table shows OLS estimation results of a panel of 54 countries from 1994m1 to 2018m12. ΔEMBI are monthly natural log returns of a country's EMBI index. HistoricalTempAnomaly is the difference between monthly temperature of a country and its 1901-1950 temperature average of the same month. In columns (1) and (2), months in the southern hemisphere are adjusted to northern hemisphere scaling (January becomes July and so on). One month or season is omitted due to multicollinearity (base category). The table shows the same results as Table 5.10, but the single terms of months or seasons are included in the depiction. Zero values indicate that the respective month is subsumed by fixed effects. Standard errors (in parentheses) are clustered at the country level, ***, ** and * indicate statistical significance at the 1%, 5% and 10% level, respectively. See Table A.5.5 for variable definitions and sources.

Table A.5.3: Robustness tests: introducing lag structure (full table)

	(1) ΔEMBI	(2) ΔEMBI	(3) ΔEMBI
DeviationAdjustedTempAnomaly	-0.275**		
	(0.125)		
L1.DeviationAdjustedTempAnomaly	0.0914		
	(0.0926)		
L2.DeviationAdjustedTempAnomaly	0.0172		
	(0.144)		
L3.DeviationAdjustedTempAnomaly	0.0130		
	(0.143)		
L4.DeviationAdjustedTempAnomaly	-0.0805		
	(0.156)		
L5.DeviationAdjustedTempAnomaly	0.00625		
	(0.0976)		
L6.DeviationAdjustedTempAnomaly	-0.172*		
	(0.103)		
L7.DeviationAdjustedTempAnomaly	-0.121		
	(0.0946)		
L8.DeviationAdjustedTempAnomaly	0.0379		
	(0.120)		
L9.DeviationAdjustedTempAnomaly	0.0750		
	(0.0779)		
L10.DeviationAdjustedTempAnomaly	-0.0195		
	(0.134)		
L11.DeviationAdjustedTempAnomaly	-0.0749		
	(0.0991)		
L12.DeviationAdjustedTempAnomaly	0.177		
	(0.125)		
HistoricalTempAnomaly		0.0474	-0.204**
		(0.109)	(0.0935)
L1.HistoricalTempAnomaly		0.104*	-0.00670
		(0.0537)	(0.0729)
L2.HistoricalTempAnomaly		0.126**	0.133*
		(0.0473)	(0.0776)
L3.HistoricalTempAnomaly		-0.0590	0.00429
		(0.0799)	(0.0774)
L4.HistoricalTempAnomaly		0.0122	0.0234
		(0.0484)	(0.0618)
L5.HistoricalTempAnomaly		-0.0636	-0.0194
		(0.0732)	(0.0961)
L6.HistoricalTempAnomaly		0.0410	-0.0998
		(0.0536)	(0.0747)
L7.HistoricalTempAnomaly		0.0267	0.104
		(0.149)	(0.0857)
L8.HistoricalTempAnomaly		-0.0635*	-0.108
		(0.0338)	(0.0777)
L9.HistoricalTempAnomaly		0.0793	-0.0396
		(0.0550)	(0.0950)
L10.HistoricalTempAnomaly		0.000139	-0.0557
		(0.0950)	(0.0810)
L11.HistoricalTempAnomaly		-0.00542	0.0320
		(0.0984)	(0.120)
L12.HistoricalTempAnomaly		0.0869	-0.0764
		(0.0849)	(0.0816)
VeryColdCountry × HistoricalTempAnomaly		-0.0360	
		(0.104)	
VeryColdCountry × L1.HistoricalTempAnomaly		-0.0699	
		(0.0536)	
VeryColdCountry × L2.HistoricalTempAnomaly		-0.0446	

	(0.0547)
VeryColdCountry $\times$ L3.HistoricalTempAnomaly	0.00932
	(0.0813)
VeryColdCountry $\times$ L4.HistoricalTempAnomaly	0.0206
	(0.0488)
VeryColdCountry $\times$ L5.HistoricalTempAnomaly	0.0442
	(0.0795)
VeryColdCountry $\times$ L6.HistoricalTempAnomaly	0.0653
	(0.0524)
VeryColdCountry $\times$ L7.HistoricalTempAnomaly	-0.0104
	(0.133)
VeryColdCountry $\times$ L8.HistoricalTempAnomaly	0.0462
	(0.0473)
VeryColdCountry $\times$ L9.HistoricalTempAnomaly	-0.0489
	(0.0691)
VeryColdCountry $\times$ L10.HistoricalTempAnomaly	-0.0980
	(0.0804)
VeryColdCountry $\times$ L11.HistoricalTempAnomaly	0.0672
	(0.0997)
VeryColdCountry $\times$ L12.HistoricalTempAnomaly	-0.122
	(0.0783)
ColdCountry (base category) $\times$ HistoricalTempAnomaly	0
	(0)
MildCountry $\times$ HistoricalTempAnomaly	-0.203
	(0.128)
MildCountry $\times$ L1.HistoricalTempAnomaly	0.0628
	(0.125)
MildCountry $\times$ L2.HistoricalTempAnomaly	-0.225**
	(0.0842)
MildCountry $\times$ L3.HistoricalTempAnomaly	0.0705
	(0.0899)
MildCountry $\times$ L4.HistoricalTempAnomaly	0.161**
	(0.0631)
MildCountry $\times$ L5.HistoricalTempAnomaly	-0.0497
	(0.0883)
MildCountry $\times$ L6.HistoricalTempAnomaly	0.0143
	(0.121)
MildCountry $\times$ L7.HistoricalTempAnomaly	-0.226
	(0.218)
MildCountry $\times$ L8.HistoricalTempAnomaly	0.0434
	(0.150)
MildCountry $\times$ L9.HistoricalTempAnomaly	-0.0139
	(0.112)
MildCountry $\times$ L10.HistoricalTempAnomaly	-0.0476
	(0.105)
MildCountry $\times$ L11.HistoricalTempAnomaly	0.0121
	(0.116)
MildCountry $\times$ L12.HistoricalTempAnomaly	-0.230
	(0.146)
WarmCountry $\times$ HistoricalTempAnomaly	-0.0649
	(0.134)
WarmCountry $\times$ L1.HistoricalTempAnomaly	-0.193**
	(0.0870)
WarmCountry $\times$ L2.HistoricalTempAnomaly	-0.0483
	(0.0657)
WarmCountry $\times$ L3.HistoricalTempAnomaly	0.0816
	(0.139)
WarmCountry $\times$ L4.HistoricalTempAnomaly	0.0401
	(0.0805)
WarmCountry $\times$ L5.HistoricalTempAnomaly	-0.0997
	(0.0820)

WarmCountry × L6.HistoricalTempAnomaly	-0.235*	
	(0.123)	
WarmCountry × L7.HistoricalTempAnomaly	-0.0614	
	(0.179)	
WarmCountry × L8.HistoricalTempAnomaly	0.163***	
	(0.0514)	
WarmCountry × L9.HistoricalTempAnomaly	-0.294**	
	(0.111)	
WarmCountry × L10.HistoricalTempAnomaly	-0.0381	
	(0.151)	
WarmCountry × L11.HistoricalTempAnomaly	0.0926	
	(0.158)	
WarmCountry × L12.HistoricalTempAnomaly	-0.000692	
	(0.140)	
VeryWarmCountry × HistoricalTempAnomaly	-0.569*	
	(0.297)	
VeryWarmCountry × L1.HistoricalTempAnomaly	-0.0300	
	(0.220)	
VeryWarmCountry × L2.HistoricalTempAnomaly	-0.0725	
	(0.256)	
VeryWarmCountry × L3.HistoricalTempAnomaly	0.00305	
	(0.220)	
VeryWarmCountry × L4.HistoricalTempAnomaly	-0.314	
	(0.206)	
VeryWarmCountry × L5.HistoricalTempAnomaly	0.248*	
	(0.139)	
VeryWarmCountry × L6.HistoricalTempAnomaly	-0.184	
	(0.216)	
VeryWarmCountry × L7.HistoricalTempAnomaly	0.0338	
	(0.192)	
VeryWarmCountry × L8.HistoricalTempAnomaly	0.0179	
	(0.265)	
VeryWarmCountry × L9.HistoricalTempAnomaly	0.0618	
	(0.128)	
VeryWarmCountry × L10.HistoricalTempAnomaly	0.148	
	(0.255)	
VeryWarmCountry × L11.HistoricalTempAnomaly	-0.0950	
	(0.177)	
VeryWarmCountry × L12.HistoricalTempAnomaly	-0.154	
	(0.204)	
RuleOfLaw		-0.00739
		(0.00642)
HistoricalTempAnomaly × RuleOfLaw		0.00369**
		(0.00139)
L1.HistoricalTempAnomaly × RuleOfLaw		0.00118
		(0.00144)
L2.HistoricalTempAnomaly × RuleOfLaw		-0.00171
		(0.00144)
L3.HistoricalTempAnomaly × RuleOfLaw		-0.00102
		(0.00130)
L4.HistoricalTempAnomaly × RuleOfLaw		0.000323
		(0.00108)
L5.HistoricalTempAnomaly × RuleOfLaw		-0.000594
		(0.00195)
L6.HistoricalTempAnomaly × RuleOfLaw		0.00290*
		(0.00153)
L7.HistoricalTempAnomaly × RuleOfLaw		-0.00241
		(0.00174)
L8.HistoricalTempAnomaly × RuleOfLaw		0.00219
		(0.00158)
L9.HistoricalTempAnomaly × RuleOfLaw		0.00121
		(0.00151)

L10.HistoricalTempAnomaly $\times$ RuleOfLaw			0.000187
			(0.00134)
L11.HistoricalTempAnomaly $\times$ RuleOfLaw			6.45e-05
			(0.00207)
L12.HistoricalTempAnomaly $\times$ RuleOfLaw			0.00113
			(0.00147)
Precipitation	0.364	0.291	0.282
	(0.404)	(0.396)	(0.391)
Observations	9,842	9,842	9,688
R-squared	0.511	0.514	0.503
Country FE	Yes	Yes	Yes
Region$\times$ MonthYear FE	Yes	Yes	Yes

This table shows robustness checks for the deviation-adjusted temperature variable (Table 5.6, column (6)), the bin-regression analyzing the warmness of countries using 5°C-intervals (Table 5.8, column (2)) and the interaction with institutional characteristics (rule of law index, Table 5.12, column (1)). The table shows the same results as Table 5.18, but the depiction for column (2) includes all other (lagged) bin-category interactions ("cold" country as base category). Estimations for columns (2) and (3) also show all lagged single terms and interactions of HistoricalTempAnomaly. Standard errors (in parentheses) are clustered at the country level, ***, ** and * indicate statistical significance at the 1%, 5% and 10% level, respectively. See Table A.5.5 for variable definitions and sources.

Table A.5.4: Robustness tests: Paris Agreement as transition shock (full table)

	(1) ΔEMBI	(2) ΔEMBI	(3) ΔEMBI	(4) ΔEMBI
HistoricalTempAnomaly	-0.00387		0.0963	-0.211**
	(0.0608)		(0.147)	(0.104)
HistoricalTempAnomaly × PostParis	-0.0363		-0.0631	0.0255
	(0.0651)		(0.159)	(0.129)
DeviationAdjustedTempAnomaly		-0.246***		
		(0.0873)		
DeviationAdjustedTempAnomaly × PostParis		0.0522		
		(0.0843)		
VeryColdCountry × HistoricalTempAnomaly			-0.0585	
			(0.135)	
VeryColdCountry × PostParis			-0.0934	
			(0.321)	
VeryColdCountry × HistoricalTempAnomaly × PostParis			0.0414	
			(0.138)	
ColdCountry (base category) × HistoricalTempAnomaly			0	
			(0)	
MildCountry × HistoricalTempAnomaly			-0.224	
			(0.153)	
MildCountry × PostParis			-0.223	
			(0.408)	
MildCountry × HistoricalTempAnomaly × PostParis			0.0957	
			(0.194)	
WarmCountry × HistoricalTempAnomaly			-0.0939	
			(0.182)	
WarmCountry × PostParis			0.229	
			(0.391)	
WarmCountry × HistoricalTempAnomaly × PostParis			-0.168	
			(0.206)	
VeryWarmCountry × HistoricalTempAnomaly			-0.608**	
			(0.257)	
VeryWarmCountry × PostParis			-0.335	
			(0.430)	
VeryWarmCountry × HistoricalTempAnomaly × PostParis			0.375	
			(0.340)	
RuleOfLaw				-0.00551
				(0.00504)
HistoricalTempAnomaly × RuleOfLaw				0.00415***
				(0.00146)
RuleOfLaw × PostParis				0.00255
				(0.00374)
HistoricalTempAnomaly × RuleOfLaw × PostParis				-0.00113
				(0.00207)
Precipitation	0.224	0.0364	0.0261	0.274
	(0.363)	(0.376)	(0.398)	(0.374)
Observations	9,957	9,957	9,957	9,688
R-squared	0.524	0.524	0.525	0.502
Country FE	Yes	Yes	Yes	Yes
Region× MonthYear FE	Yes	Yes	Yes	Yes

This table shows robustness checks for the temperature anomaly measure (Table 5.6, column (3)), the deviation-adjusted temperature variable (Table 5.6, column (6)), the bin-regression analyzing the warmness of countries using 5°C-intervals (Table 5.8, column (2)) and the interaction with institutional characteristics (rule of law index, Table 5.12, column (1)). PostParis is a dummy with value 1 after the Paris Climate Agreement in December 2015. Estimation in column (3) also shows all other bin categories (cold as base category) and respective interactions that were not shown in Table 5.24. Standard errors (in parentheses) are clustered at the country level, ***, ** and * indicate statistical significance at the 1%, 5% and 10% level, respectively. See Table A.5.5 for variable definitions and sources.

Table A.5.5: Description and sources of variables

Variable	Description	Source
Variables in Baseline Regression (Section 5.4)		
ΔEMBI	Monthly change in natural logarithm of Emerging Market Bond Index (Global) (winsorized at 1st and 99th percentile)	J.P. Morgan
Historical Temperature Anomaly (Historical TempAnomaly)	Difference between monthly temperature of a country and its 1901-1950 temperature average of the same month	Climatic Research Unit, see Harris et al. (2020)
Deviation-Adjusted Temperature Anomaly (DeviationAdjusted-TempAnomaly)	HistoricalTempAnomaly divided by a country's 1901-1950 standard deviation of monthly temperature	Climatic Research Unit, see Harris et al. (2020)
Precipitation	Precipitation in units of 1000 mm per month	Climatic Research Unit, see Harris et al. (2020)
ΔVIX	Monthly first difference in VIX volatility index (winsorized at 1st and 99th percentile)	CBOE
ΔUS-CorporateRisk Premium	Monthly first difference in spread between the S&P US high yield corporate bond index and the corresponding investment grade index (winsorized at 1st and 99th percentile)	S&P
ΔUS-10-YearTreasury Yield	Monthly first difference in the yield of the 10-year US Treasury bond (winsorized at 1st and 99th percentile)	Datastream
ΔUS-TermSpread	Monthly first difference in spread between 10-year US Treasury yield and 3-month US T-Bill yield (winsorized at 1st and 99th percentile)	Datastream, Federal Reserve
ΔGlobalGovernment BondIndex	Monthly change in natural logarithm of Bank Of America Merrill Lynch Global Government Index (winsorized at 1st and 99th percentile)	Merrill Lynch
Variables in Interaction and Bin Regressions (Section 5.5)		
Very cold, cold, mild, warm, very warm country (percentile)	Countries are grouped into a bin according to percentile distribution of average annual temperature (1901-2018), 1st-20th (very cold), 21st-40th (cold) percentile and so on	
Very cold, cold, mild, warm, very warm country (5°C-interval)	Countries are grouped into a bin according to 5°C-intervals $\leq 10°$C (very cold), > 10 & $\leq 15°$C (cold), > 15 & $\leq 20°$C (mild), > 20 & $\leq 25°$C (warm), $> 25°$C (very warm)	
Very low, low, normal, high, very high temperature standard deviation (Temp-StdDev)	Countries are grouped into a bin according to percentile distribution of monthly temperature standard deviation (1901-2018), 1st-20th (very low), 21st-40th (low) percentile and so on	
Spring	Dummy, 1 in months March-May for northern and September-November for southern hemisphere countries	
Summer	Dummy, 1 in months June-August for northern and December-February for southern hemisphere countries	
Autumn	Dummy, 1 in months September-November for northern and March-May for southern hemisphere countries	
Winter	Dummy, 1 in months December-February for northern and June-August for southern hemisphere countries	
Agriculture to GDP	Value added of agriculture (% of gross domestic product)	World Bank

Manufacturing to GDP	Value added of manufacturing (% of gross domestic product)	World Bank
Services to GDP	Value added of services (% of gross domestic product)	World Bank
Resource Rents to GDP	Sum of oil rents, natural gas rents, coal rents (hard and soft), mineral rents, and forest rents (% of gross domestic product)	World Bank
Rule of Law	Rule of law rank (the extend of which agents have confidence in and abide by the rules of society; linearly interpolated)	World Bank
Control of Corruption	Control of corruption rank (the extent to which public power is exercised for private gain, including both petty and grand forms of corruption, as well as "capture" of the state by elites and private interests; linearly interpolated)	World Bank
Civil Liberties	Countries and territories with a rating of 1 enjoy a wide range of civil liberties. Countries and territories with a rating of 7 have few or no civil liberties	Freedom House
Political Rights	Countries and territories with a rating of 1 enjoy a wide range of political rights, including free and fair elections. Countries and territories with a rating of 7 have few or no political rights	Freedom House
Income Redistribution	Absolute income redistribution (market income inequality minus net-income inequality)	Solt (2019)
Polity2	Unified polity scale that ranges from +10 (strongly democratic) to -10 (strongly autocratic)	Center for Systemic Peace
Democratic Government	Scale that ranges from 0 (not democratic) to +10 (strongly democratic) government (standardized authority codes set to missing)	Center for Systemic Peace
Authoritarian Government	Scale that ranges from 0 (not authoritarian) to +10 (strongly authoritarian) government (standardized authority codes set to missing)	Center for Systemic Peace
ND-GAIN	Notre Dame Global Adaption Index; ND-GAIN brings together over 74 variables to form 45 core indicators to measure vulnerability and readiness to climate change	Notre Dame Global Adaption Initiative
Readiness Index	Readiness component of ND-GAIN; measures readiness by considering a country's ability to leverage investments to climate adaptation actions	Notre Dame Global Adaption Initiative
Vulnerability Index	Vulnerability component of ND-GAIN; measures propensity or predisposition of human societies to be negatively impacted by climate hazards	Notre Dame Global Adaption Initiative
GDP per Capita	Gross domestic product per capita in constant 2010-US-dollar prices	World Bank

Variables in Robustness Tests (Section 5.6)

ΔEMBI Spread	Monthly first difference in Emerging Market Bond Spread (Global) (winsorized at 1st and 99th percentile)	J.P. Morgan
ΔCDS Spread	Monthly first difference in sovereign CDS Spread (winsorized at 1st and 99th percentile)	Thomson Reuters CDS
Historical Deviation Anomaly	Difference between a country's 12-month rolling temperature standard deviation and its 1901-1950 standard deviation of temperature	Climatic Research Unit, see Harris et al. (2020)
Debt to GDP	General gross government debt (% of gross domestic product)	Oxford Economics
Years since last Sovereign Debt Restructuring	Number of years since last sovereign debt restructuring. Value of 36 if no sovereign debt restructuring took place	Laeven & Valencia (2018)

Primary Net Lending to GDP	Government primary net lending/borrowing (% of gross domestic product)	IMF Fiscal Monitor
Post Paris	Dummy that is 1 after Paris Agreement (December 2015)	
Agriculture Land Share	Agricultural land (% of total land area)	World Bank
Agriculture Employment Share	Employment in agriculture (% of total employment)	World Bank
Industrial Employment Share	Employment in industry (% of total employment)	World Bank
Services Employment Share	Employment in services (% of total employment)	World Bank

Further data used		
Natural Disasters	Date of drought, earthquake, epidemic, heat wave, flood, impact, insect infestation, landslide, mass movement, storm, volcanic activity, wildfire (total deaths, damage and affected people for certain disasters)	International Disaster Database
Stock Returns	Natural log returns of stock market index	MSCI, S&P
GDP Growth	Quarterly natural log change of GDP in constant, seasonally-adjusted 2015 US-Dollar prices	Oxford Economics
Accumulated CO_2 Emissions	Accumulated CO_2 emissions of every country and the world since 1751	Global Carbon Project, retrieved via ourworldindata.org
Population	Total population of every country and the world in 2017	World Bank

References to Chapter 5

Acemoglu, D., Johnson, S. & Robinson, J. A. (2002), 'Reversal of fortune: Geography and institutions in the making of the modern world income distribution', *The Quarterly Journal of Economics* **117**(4), 1231–1294.

Addoum, J. M., Ng, D. T. & Ortiz-Bobea, A. (2020), 'Temperature shocks and establishment sales', *The Review of Financial Studies* **33**(3), 1331–1366.

Adhvaryuy, A., Kalaz, N. & Nyshadham, N. (2020), 'The light and the heat: Productivity co-benefits of energy-saving technology', *Review of Economics and Statistics* **102**(4), 1–14.

Angrist, J. D. & Pischke, J.-S. (2008), *Mostly harmless econometrics: An empiricist's companion*, Princeton university press.

Auffhammer, M. & Schlenker, W. (2014), 'Empirical studies on agricultural impacts and adaptation', *Energy Economics* **46**, 555–561.

Bank of England (2018), 'Transition in thinking: The impact of climate change on the UK banking sector'.

Bansal, R., Kiku, D. & Ochoa, M. (2016), 'Price of long-run temperature shifts in capital markets'. NBER Working Paper No. 22529.

Barrios, S., Bertinelli, L. & Strobl, E. (2010), 'Trends in rainfall and economic growth in Africa: A neglected cause of the African growth tragedy', *The Review of Economics and Statistics* **92**(2), 350–366.

Bathiany, S., Dakos, V., Scheffer, M. & Lenton, T. M. (2018), 'Climate models predict increasing temperature variability in poor countries', *Science Advances* **4**(5, eaar5809), 1–10.

Bloomberg (2020), 'Australia sued for not disclosing climate risk in sovereign debt'. `https://www.bloomberg.com/amp/news/articles/2020-07-22/australia-sued-for-not-disclosing-climate-risk-in-sovereign-debt?__twitter_impression=true`.

Bolton, P., Despres, M., Pereira da Silva, L. A., Samama, F. & Svartzman, R. (2020), 'The green swan - Central banking and financial stability in the age of climate change'. Bank for International Settlements.

Bolton, P. & Kacperczyk, M. (2020), 'Do investors care about carbon risk?'. NBER Working Paper No. 26968.

Boston Common Asset Management (2018), 'Banking on a low-carbon future: Are the world's largest banks stepping up to the risks & opportunities of climate change?'. `http://news.bostoncommonasset.com/wp-content/uploads/2018/02/Banking-on-a-Low-Carbon-Future-2018-02.pdf`.

Bun, M. J. & Harrison, T. D. (2019), 'OLS and IV estimation of regression models including endogenous interaction terms', *Econometric Reviews* **38**(7), 814–827.

Burke, M. B., Miguel, E., Satyanath, S., Dykema, J. A. & Lobell, D. B. (2009), 'Warming increases the risk of civil war in Africa', *Proceedings of the National Academy of Sciences* **106**(49), 20670–20674.

Burke, M., Hsiang, S. M. & Miguel, E. (2015), 'Global non-linear effect of temperature on economic production', *Nature* **527**(7577), 235–239.

Cachon, G. P., Gallino, S. & Olivares, M. (2012), 'Severe weather and automobile assembly productivity'. Columbia Business School Research Paper No 12/37.

Delis, M., De Greiff, K. & Ongena, S. (2018), 'Being stranded on the carbon bubble? Climate policy risk and the pricing of bank loans'. SFI Research Paper 8-10, University of Zurich.

Dell, M., Jones, B. F. & Olken, B. A. (2012), 'Temperature shocks and economic growth: Evidence from the last half century', *American Economic Journal: Macroeconomics* **4**(3), 66–95.

Dell, M., Jones, B. F. & Olken, B. A. (2014), 'What do we learn from the weather? The new climate-economy literature', *Journal of Economic Literature* **52**(3), 740–98.

Deryugina, T. & Hsiang, S. M. (2014), 'Does the environment still matter? Daily temperature and income in the United States'. NBER Working Paper No. 20750.

Easterly, W. & Levine, R. (2003), 'Tropics, germs, and crops: How endowments influence economic development', *Journal of Monetary Economics* **50**(1), 3–39.

ECB (2019), 'Financial stability review May 2019'.

Eichler, S. (2014), 'The political determinants of sovereign bond yield spreads', *Journal of International Money and Finance* **46**, 82–103.

ESRB (2016), 'Too late, too sudden: Transition to a low-carbon economy and systemic risk'. Reports of the Advisory Scientific Committee No 6.

Felbermayr, G. & Gröschl, J. (2014), 'Naturally negative: The growth effects of natural disasters', *Journal of Development Economics* **111**, 92–106.

Financial Stability Institute (2019), 'Turning up the heat - Climate risk assessment in the insurance sector'. FSI Insights on policy implementation No 20.

Germanwatch (2019), 'Global climate risk index 2020'. `https://germanwatch.org/sites/ germanwatch.org/files/20-2-01e%20Global%20Climate%20Risk%20Index%202020_ 13.pdf`.

Graff Zivin, J. & Neidell, M. (2014), 'Temperature and the allocation of time: Implications for climate change', *Journal of Labor Economics* **32**(1), 1–26.

Gupta, M. S., Mati, A. & Baldacci, M. E. (2008), 'Is it (still) mostly fiscal? Determinants of sovereign spreads in emerging markets'. IMF Working Paper 08/259.

Harris, I., Osborn, T. J., Jones, P. & Lister, D. (2020), 'Version 4 of the CRU TS monthly high-resolution gridded multivariate climate dataset', *Scientific Data* **7**(1), 1–18.

Heal, G. & Park, J. (2014), 'Feeling the heat: Temperature, physiology & the wealth of nations'. NBER Working Paper No. 19725.

Hilscher, J. & Nosbusch, Y. (2010), 'Determinants of sovereign risk: Macroeconomic fundamentals and the pricing of sovereign debt', *Review of Finance* **14**(2), 235–262.

Hsiang, S. M., Burke, M. & Miguel, E. (2013), 'Quantifying the influence of climate on human conflict', *Science* **341**(6151), 1235367.

IPCC (2018), 'Global warming of 1.5°C. An IPCC Special Report on the impacts of global warming of 1.5°C above pre-industrial levels and related global greenhouse gas emission pathways, in the context of strengthening the global response to the threat of climate change, sustainable development, and efforts to eradicate poverty (summary for policymakers)', *Intergovernmental Panel on Climate Change* .

Kling, G., Lo, Y. C., Murinde, V. & Volz, U. (2018), 'Climate vulnerability and the cost of debt'. Available at SSRN 3198093.

Koetter, M., Noth, F. & Rehbein, O. (2020), 'Borrowers under water! Rare disasters, regional banks, and recovery lending', *Journal of Financial Intermediation* **43**, 100811.

Laeven, L. & Valencia, F. (2018), 'Systemic banking crises revisited'. IMF Working Paper 18/206.

Moody's (2016), 'Environmental risks - sovereigns: How Moody's assesses the physical effects of climate change on sovereign issuers'. `https://www.moodys.com/research/Moodys-sets-out-approach-to-assessing-the-credit-impact-of--PR_357629`.

Nizalova, O. Y. & Murtazashvili, I. (2016), 'Exogenous treatment and endogenous factors: Vanishing of omitted variable bias on the interaction term', *Journal of Econometric Methods* **5**(1), 71–77.

Painter, M. (2020), 'An inconvenient cost: The effects of climate change on municipal bonds', *Journal of Financial Economics* **135**(2), 468–482.

Pankratz, N. & Schiller, C. (2019), 'Climate change and adaptation in global supply-chain networks'. Available at SSRN 3475416.

Ripple, W. J., Wolf, C., Newsome, T. M., Barnard, P. & Moomaw, W. R. (2019), 'World scientists' warning of a climate emergency', *BioScience* .

Sachs, J. D. (2001), 'Tropical underdevelopment'. NBER Working Paper No. 8119.

Solt, F. (2019), 'Measuring income inequality across countries and over time: The standardized world income inequality database'. SWIID Version 8.2.

Somanathan, E., Somanathan, R., Sudarshan, A., Tewari, M. et al. (2018), 'The impact of temperature on productivity and labor supply: Evidence from Indian manufacturing'. Becker Friedman Institute Working Paper No. 2018-69.

Zhang, P., Deschenes, O., Meng, K. & Zhang, J. (2018), 'Temperature effects on productivity and factor reallocation: Evidence from a half million Chinese manufacturing plants', *Journal of Environmental Economics and Management* **88**, 1–17.